FRANK FRASER DARLING is remember
and a founding father of the green n
responsible for a stream of popular
with his environment. Not beloved by the scientific establishment, he felt he
had to move to East Africa and the USA to have his ideas put into practice.
Pelican in the Wilderness and the Reith Lectures of 1969 (*Wilderness and Plenty*) brought the environment into focus for the first time and Fraser Darling belatedly found himself flavour of the month. Anthony Crosland was made the first Secretary of State for the Environment and set up a Royal Commission on the Environment. Fraser Darling was appointed as one of its founder members and remained on it until his final retirement from public business in 1973. One of Harold Wilson's last acts in the week he lost the 1970 Election was to propose Fraser Darling for a Knighthood.

A Herd of Red Deer was Fraser Darling's first substantial publication. It was written as the outcome of the first major study of a large mammal based on long periods of observation in the wild and is almost a modern classic of writing on natural history, comparable to Gilbert White's (1720–1793) *Natural History of Selborne*. Much more than a description of how animals behave in a wild landscape – the most savage mountain area in the British Isles – Fraser Darling writes easily and with sympathy for these fine animals.

Like many of his generation, WALTER STEPHEN was heavily influenced in his youth by Frank Fraser Darling and his writings. From him he learned to love the wild parts of Scotland, not just for their beauty and their romantic past, but as the end-product of a series of complex relationships between structure and climate, plants and animals, conservation and exploitation, locals and incomers. Like Fraser Darling he has accumulated many man-hours in the wild places – although without Fraser Darling's dedication and keen observation. Professionally, over the years he has tried, in many ways, to help others to follow Fraser Darling in adding understanding and concern for the environment to the passive enjoyment of scenery.

At this time, with an enhanced sense of the importance of Scottish heritage and an increased awareness of the environment and its challenges, Stephen feels that *A Herd of Red Deer* has much to offer the wise reader.

Praise for Frank Fraser Darling

He had the broad picture, the grand sweep, he was a prophet and a guru.
PALMER NEWBOULD, Emeritus Professor of Environmental Science, University of Ulster

Frank Fraser Darling had a gift of writing scientifically in a biographical way.
JOHN MORTON BOYD, Late Director, Nature Conservancy Council in Scotland

Frank Fraser Darling's landmark study provided the foundation for and stimulated similar studies throughout the world.
PROFESSOR TIM CLUTTON-BROCK, Prince Philip Professor of Ecology and Evolutionary Biology, University of Cambridge

This book has warmth and personality and an infectious appreciation of the good things of life.
V WYNNE-EDWARDS, *The Scottish Naturalist*, 1948

A book of rare value to sportsmen and naturalists. No other gives a more intimate or accurate tale of the doings of the deer where the grey fox lurks and the eagle hunts.
BOOKS OF THE DAY, *The Scotsman*, 13 January 1938

A Herd of Red Deer
A Study in Animal Behaviour

F. FRASER DARLING

Introduced and edited by Walter Stephen

Luath Press Limited
EDINBURGH
www.luath.co.uk

This edition reprinted by Luath Press by arrangement with Oxford University Press. First Published by Oxford University Press in 1937. © Oxford University Press.

This edition 2008
Reprinted 2011

ISBN: 978-1-906307-42-4

The publishers acknowledge the support of

 and

towards the publication of this volume.

The paper used in this book is acid-free, neutral-sized and recyclable. It is made from low-chlorine pulps produced in a low energy, low emissions manner from renewable forests.

Printed and bound by
Bell & Bain Ltd., Glasgow

Typeset in 11 point Sabon
by 3btype.com

Contents

Foreword	xi
Acknowledgements	xii
Introduction	xiii
Frank Fraser Darling 1903–1979, Ecologist,	
Conservationist, Prophet	xvi
A Herd of Red Deer	xxvi
Relevance Today	xxxiii
Fraser Darling and the Environmental Movement	xxxvii
CHAPTER ONE The Country	1
CHAPTER TWO Technique and Personal Reactions	18
Equipment	18
Stalking	20
CHAPTER THREE Territory and population	27
Winter Territory	29
Summer Territory	29
Breeding or Rutting Territories	30
CHAPTER FOUR Territory And Population (contd.)	47
Deer Paths, Wallows, and Rubbing Trees	58
CHAPTER FIVE Territory and the Social System	64
The Hind Group	65
Stag Companies	72
The Harem	78
CHAPTER SIX Some Social Factors and Comparisons	80
The Voice	80
Play	82
General Remarks on Sociality;	
and Comparisons	87
Conclusion	91

CHAPTER SEVEN	Movement: The Influence Of Weather	95
	Meteorological Influences on Movement	97
	Temperature	97
	Barometric Pressure	108
	Humidity	108
	Wind	117
	Precipitation	118
	Light and Darkness	123
CHAPTER EIGHT	Movement: The Influence of Insects and Food Supply	126
	Biological Influences on Movement: Arthropodan Sources of Disturbance	126
	Helminth Parasites	140
	Vegetation	141
	Plant Associations and Movement of Deer	143
CHAPTER NINE	Reproduction	150
	The Sexual Psycho-Physiology of the Stag	150
	The Antlers	153
	Behaviour of the Stag from the Casting of the Antlers until the Onset of the Rut	157
	The Role of Temperature in the Development and Maintenance of the Rut	162
	Movement of Stags during the Rut	164
	The Course of the Rut	166
	The Hind at Calving Time	176
CHAPTER TEN	The Senses	179
	Smell	180
	Hearing	184
	Sight	185
	Touch	189
	Taste	189
CHAPTER ELEVEN	Conclusion	191
Glossary		197
Bibliography		199
Index		207

Illustrations – Colour Plates

PLATE 1A	Sgurr Fiona and Coire Toll an Lochain *Walter Stephen*
PLATE 1B	Carn na Carnach – the favourite ground of the deer *Walter Stephen*
PLATE 2A	A herd of red deer in winter © *Deer Commission for Scotland*
PLATE 2B	Stag feeding on heather *Neil McIntyre*
PLATE 3A	Hind keeping watch on hill © *Deer Commission for Scotland*
PLATE 3B	Stags boxing in early May *Neil McIntyre*
PLATE 4A	June – Hind and calf *Neil McIntyre*
PLATE 4B	Summer – Stag in velvet *Neil McIntyre*
PLATE 5A	August – Stag rubbing tree *Neil McIntyre*
PLATE 5B	Deer wallow *Laurie Campbell*
PLATE 6A	Stag roaring during the autumn rut *Neil McIntyre*
PLATE 6B	The rut – Stag thrashing its head in peat bog *Neil McIntyre*
PLATE 7A	The rut – Stags sparring *Neil McIntyre*
PLATE 7B	The rut – Stag scenting hind *Neil McIntyre*
PLATE 8A	Red deer stag and hinds in a quiet moment *Laurie Campbell*
PLATE 8B	Autumn stags in Strath Dearn *Laurie Campbell*

Illustrations – Black and White

Frontispiece	Red stags in velvet	x
	Sgurr Fiona and the Toll Lochan corrie	7

The pine wood, Dundonnell	54
Beinn Dearg Mhor	77
Looking up the Gruinard River	109
Stag wallow in Glen Muic	138
Hind wallow in Carn na Carnach	138
A rubbing tree	156
A young stag in 'velvet'	156
Hinds grazing	187
They heard the click of the camera	187

Illustrations – Figures

FIG 1	Roderick and Flora watch the deer *Sandy the Red Deer*, Oxford University Press	xxiii
FIG 2	The herd in winter *Sandy the Red Deer*, Oxford University Press	xxiv
FIG 3	Diagram of northern portion of Carn na Carnach hind territory	69
FIG 4	Movement in red deer	96
FIG 5	Monthly average maximum (shade) and minimum temperatures taken at Brae House, Dundonnell – altitude 150 feet	98
FIG 6	Differences between average monthly maximum and minimum temperatures taken at Brae House, Dundonnell – altitude 150 feet	99
FIG 7	Thermograph and Hygrograph readings 7–9 February 1936	111
FIG 8	Thermograph and Hygrograph readings 18–20 October 1935	112
FIG 9	Diagram to show irritability to movement	

	by disturbance in relation to humidity of the atmosphere	113
FIG 10	Thermograph and Hygrograph readings 13 June 1934	115
FIG 11	Thermograph and Hygrograph readings 17–22 December 1935	116
FIG 12	Growth of antlers and maturation of gonads	161

Illustrations – Maps

MAP 1	Geological sketch-map of the terrain	4
MAP 2	Distribution of red deer in Scotland	26
MAP 3	Rutting territories	31
MAP 4	Hind territories	32
MAP 5	Stag territories	32
MAP 6	Map of parts of the forests of Dundonnell, Gruinard and Letterewe	46
MAP 7	Deer paths and wallows	61

Red Stags in 'Velvet'
The stags use their fore-feet in quarrelling when the antlers are soft.

Foreword

THE REPUBLISHING OF *A Herd of Red Deer* 70 years after it first appeared is for us an exciting event. These were the years of Alasdair's childhood, living in the isolated Brae House in Dundonnell, looking across the glen to that magnificent mountain, An Teallach. They were happy times and there are many good memories including climbing An Teallach with snow still in the corries. Our father was totally happy and absorbed in this groundbreaking study of animals in their natural environment, a method of study that is commonplace now but then was considered to be unscientific.

Many of the issues discussed in the book remain pertinent today. In the absence of effective predators, red deer increase at a rate faster than that of culling with consequent pressure on the Highlands habitat. There are fierce arguments over the correct balance between economic interests such as stalking and farming, road safety and conservation of the environment. We very much hope that this republication serves as a constructive contribution to that debate as well as providing a still relevant insight into Britain's largest wilderness area, a landscape that arouses passion in all who experience it.

It is with gratitude that we welcome Walter Stephen's inspiration to republish this work and even more to bring it to fruition. His Introduction provides a context in which our father's work was undertaken and Dr Stephen has a sensitive understanding of his thinking and way of working. From his own experience of climbing in this isolated part of Wester Ross he reveals his admiration of our father's ability to live on the hill for days at a time.

Alasdair Fraser-Darling
Richard Fraser Darling

Acknowledgements

The editor makes no claim whatsoever to expertise in the areas of knowledge and understanding in which Frank Fraser Darling was an innovator and an inspiration. In bringing into the public domain again the pioneering work *A Herd of Red Deer* he recalls his own lifetime absorption in the wild places of Scotland – especially Wester Ross – and the inspiration and understanding he drew from Fraser Darling's writings.

But he can only contribute some low-grade reminiscences of one who was around at the time and this new edition would have been a poor thing had it been his work alone.

Many experts in the world of ecology have been similarly touched by Fraser Darling's life and achievements and the editor is grateful and pleased that so many have committed themselves to giving him expert advice. It tells us something about Fraser Darling that, after such a passage of time, there should be a substantial pool of people who remember him, respect him or feel that he has influenced them.

Principal among those who have materially supported me in my attempt to make *A Herd of Red Deer* accessible to the 21st century public are his sons, Dr Alasdair Fraser-Darling and Richard Fraser Darling, Professor Tim Clutton-Brock, Prince Philip Professor of Ecology and Evolutionary Biology, University of Cambridge, Dr Aubrey Manning, Emeritus Professor of Natural History, University of Edinburgh, Dr Palmer Newbould, Emeritus Professor of Environmental Science, University of Ulster, Dr Des Thompson, Principal Uplands Advisor, Scottish Natural Heritage and Alastair MacGugan, Deer Commission for Scotland.

Scottish Natural Heritage have grant-aided the publication and showed a general commitment to the project. The Deer Commission for Scotland gave access to their photographic library and allowed the publication of their images free. Oxford University Press permitted the publication of two sketches from *Sandy the Red Deer* without charging a fee.

Introduction

WHEN THE EDITOR first began work on *A Herd of Red Deer* a frequently asked question was: 'Why choose that book?' Fraser Darling was responsible (with occasional collaboration) for some 60 papers and 20 books, some of which are better known now than *A Herd of Red Deer* and sold more copies when they were first published. Yet the book was a milestone, one of the first along the road to our present state of knowledge about the natural world and how it is affected by the works of man.

A Herd of Red Deer was the result of work carried on in an early stage of his career and was probably his greatest research achievement. It is based entirely on painstaking field observations and established the importance of social behaviour in the seasonal movements, grazing patterns and breeding of red deer stags and hinds. At the time an eminent reviewer wrote:

> Without any doubt Fraser Darling has set a new standard for field study of large mammals.

For me, this is not strong enough. This book is revolutionary in that it describes, for the first time, how a scientist left the laboratory – with its artificial conditions – and the library in order to go out into the habitat to be in close observation of the wild animals there. Living as we do in an age where intrusive cameras are everywhere it is probably difficult to grasp how wonderful this approach to research was in Fraser Darling's time.

Fraser Darling describes his findings and searches for explanations, but in no dry-as-dust way. He carries the reader along. In addition, as the reviewer quoted above said:

> There is a strong element of aesthetic enjoyment running through as well as a spontaneous sympathy with wild animals and a desire to communicate this to others.

Another obvious question was: 'Why did he choose the red deer to study?' The red deer (*Cervus elephas*) is our largest land mammal. Although not unique to Scotland, 28 per cent of the red deer population of Europe is in Scotland, and is unusual in that the deer has had to evolve from an animal of the forests to one that can exist on bare moorland. In turn this means that it can often be seen from main roads, especially in bad weather when the deer are forced down from the higher hills.

The red deer is good to eat and has been hunted for food since the first settlers came to what is now Scotland. The red deer stag is a handsome beast and, because deer are (rightly) very wary of human presence, a worthy challenge for the sportsman. Unlike forested continental Europe, where the deer are sometimes driven past the guns, killing deer in Scotland requires skilled fieldcraft as well as good shooting.

'The stag at eve had drunk his fill', wrote Sir Walter Scott, as *The Lady of the Lake* got under way, soon to generate a great flow of tourists to the Highlands. After most of the original forest cover had been removed, big landowners cleared their estates. Selective over-grazing having turned them into man-made deserts (a process repeatedly described by Fraser Darling) they were turned into deer forests – with few trees.

Prince Albert came, and saw, and killed a great many stags. When *Leaves from the Journal of Our Life in the Highlands* was published on New Year's Day 1868 neither the Queen's nor the Prince Consort's names appeared on the title pages, nor did the royal cipher or coat of arms appear on the covers. Instead there was a small pair of antlers back and front, interlaced to form the initials V and A on a moss-green ground. Landseer visited Glen Feshie and painted *The Stag at Bay*, the ultimate Victorian icon, since much ridiculed and plagiarised. No-one who saw the recent film of *The Queen* could fail to miss the symbolism of the great stag and the degradation of its fate.

The largest by far, if not the best, picture in the National Gallery of Scotland was painted by the American Benjamin West in 1786 and was commissioned by the last Lord Seaforth. 'West's Scottish masterpiece' – *Alexander III of Scotland rescued from the fury of a stag by the intrepidity of Colin Fitzgerald* – shows the poor stag, menaced by at least half a dozen ravening dogs and armed courtiers. The unhorsed king is on the ground but the stag has been seized by the antlers by a heroic Fitzgerald who, garbed in an odd Highland/Roman costume, is about to deliver the mortal thrust from his upraised spear.

This is the founding legend of the MacKenzies. The King's gratitude started the rise of the MacKenzies of Seaforth and the stag's antlers appeared on their coats of arms, and later on regimental colours, buttons, badges and the like. *Cabar feidh* means the antlers of the stag and describes the way Highland dancers' arms are held aloft. Highland dancing is, of course, a celebration of male virility and the leaping represents the leap of the stag at the moment of copulation. *Cabar feidh* – one of the most stirring of Highland tunes – became the regimental charge of the Earl of Seaforth's Highland Regiment in 1778, functioning as a duty tune of the Seaforths till the amalgamations began in 1961.

Red deer have interesting social patterns and move around a great deal. During the rut males and females are together, at other seasons stags and hinds seem to be quite separate. Sometimes the deer are high on the hill, at others they come right down into the glens. Sleep is often impossible at the time of the rut. Prior to Fraser Darling there was a great deal of lore about deer, mainly anecdotal and mainly centred on the chase, stalking and killing, but also based on the natural history observations of such as Seton Gordon.

The red deer is a big handsome mammal with an interesting lifestyle. Although wild, it is heavily influenced by man. It lives in some of the most beautiful but most hostile habitats in Britain, where field observation becomes an adventure. As Fraser Darling says:

> If we are to watch one of the higher animals ... the subject
> for study should exhibit marked reactions; and ... it is better
> for us that the animal should live above ground.

Given the choice between the vole in the meadow and the deer on the slopes of An Teallach, the answer becomes inevitable for a mind like Fraser Darling's – and he was already an expert on hoofed mammals.

Palmer Newbould asks the basic question about Frank Fraser Darling – 'How did a man with no first degree, little academic training and little serious ecological research achieve such a high status in the international world of conservation?'

Fraser Darling started off well. His personality was a mixture of charm and authority. He seems to have known the right people at crucial points of his career. He had the self-confidence of having worked with his hands and, later, of having been the first to study large animals by spending long periods in close proximity to them.

As a naturalist he was a keen observer and could empathise with his subjects. (Remember that Gilbert White of Selborne was a parish minister.) When it came to communication he had the gift of writing well, making his subjects accessible, even exciting, without 'dumbing-down' as we would now say. It would be interesting to know how many youngsters in the post-war years went in for ecological or rural careers because of Fraser Darling.

As Palmer Newbould says: 'He had the broad picture, the grand sweep, he was a prophet and a guru.' As a pioneer of ecology he did not have to concern himself too much with painstaking, nit-picking micro-studies. From his observations in the field he was able to move on quickly to develop policies and management, to combine plant, animal and human ecology. He was fortunate to move on latterly to the United States where they had a much greater awareness of the damage man's impact on the environment could have and which he had borne witness to in his *West Highland Survey*. Over there they understood and listened to him while in this country nobody wanted to know.

Frank Fraser Darling 1903–1979, Ecologist, Conservationist, Prophet

FRASER DARLING WAS brought up in Sheffield and left school at 15 to work on a farm in the Derbyshire High Peak. From this valuable hands-on experience he went on to a Diploma Course at East

Midlands Agricultural College. At Edinburgh he did a PH.D. on the Blackface sheep and was attached to the Imperial Bureau of Animal Genetics.

William Hesketh Lever, Lord Leverhulme, known pejoratively as 'The Soapman' by the islanders he tried to help, had died in 1925 but there were still Leverhulme Fellowships for those who sought to improve the Highlands and Islands. Thus Arthur Geddes, younger son of Sir Patrick, surveyed Lewis and, in 1933, Fraser Darling began to work on red deer in the West Highlands, culminating in 1937 with *A Herd of Red Deer*. Palmer Newbould considers this his greatest research achievement, based entirely on painstaking field observations and establishing the importance of social behaviour in the seasonal movements, grazing patterns and breeding of red deer stags and hinds.

The detail was important but more significant was the method. Fraser Darling was among the first of his kind to break out from the laboratory and spend long periods in close observation of the subjects in the wild. He did not live with the deer, because no-one can live with a herd of red deer, but he spent substantial periods 'on the hill', often returning to base only at the weekend.

Fraser Darling's stance was that the great bulk of papers on animal behaviour in his time lifted the organism from its normal environment and placed it in a set of artificial conditions. Often the results were not valid for interpretation of representative behaviour. For him, preliminary studies of animals in their natural surroundings should be the initial steps to any experimental approach – and these were then of rare occurrence. Studies such as his took much time and patience and frequently isolated the observer from intellectual contact with his fellows. They were not tasks for the laboratory and daily common-room discussion. One result of this was that he never quite fitted in to the scientific establishment.

The anonymous reviewer in *The Scotsman* of 13 January 1938 concentrated on telling readers what was in the book, rather than analysing it. But he was respectful and generous, although he underestimated the target audience when he suggested it was: 'a book of rare value to sportsmen and naturalists.'

A Herd of Red Deer was essentially a serious, scholarly account of a research project, although there were subjective descriptions and deductions. *A Naturalist on Rona*: *Essays of a Biologist in Isolation* (1939) carried on the same kind of work, this time focussing on seals. By 1940, when *Island Years* was published, Fraser Darling had given many factual accounts of his work in public, to find that the audience response was: 'But what *we* want to know is how *you* lived, please.'

So *Island Years* is, in Fraser Darling's words: 'a chapter of experience, three years of three people's lives'. This is a narrative tale, starting:

> When we studied the behaviour of the red deer and the Atlantic seals we lived as near their life as we could

and concluding:

> And still we are among the islands, making a little farm from land that was derelict, seeing new things and seeing old things again in a new beauty.

It is interesting for us to note, from our acquaintance with reality television, the interest of the press in the Fraser Darlings and their voluntary exile in a succession of islands. Fraser Darling was at pains to point out in print his unhappiness about the media coverage – with the honourable exception of the *Glasgow Herald*.
Island Farm' (1943) was:

> A shriving experience starting from scratch and reaching a peasant level of comfort in three and a half years.

A popular and absorbing narrative, it was much more than that. Fraser Darling was able to use his experience on Tanera Mor, in the Summer Isles, as evidence for his post-war proselytising work. While others could plan from the study or the drawing table, Fraser Darling had been through the mill of living like a pioneer crofter – but with the benefit of scientific reflection:

> We have reached certain convictions – one is that one family is too small a unit to live alone on a small island. Second, there is nothing intrinsic in the peasant life to prevent corporate and cultural development.

INTRODUCTION

In the immediate post-Second World War years there was great hunger for knowledge from those who had served in the forces and now felt they had to make up for lost time. For others the war had meant, quite simply, fewer books published. A new, better, world had to be created. Hence the popularity of such titles as *Science for the Citizen* and *Mathematics for the Million*. Collins the publishers contribution was the *New Naturalist* series.

These were handsome books, authoritative but accessible, with good illustrations, many in colour – which was a delight after the years of austerity. Many of the subjects were important, with leading writers in each field. Thus there was Manley on *Climate and the British Scene* (1952), Steers on *The Sea Coast* (1945), Richard Fitter on *London's Natural History* (1945). Dudley Stamp, author of over 100 titles and the dynamo behind the Land Utilisation Survey of Britain in the 1930s, contributed *Britain's Structure and Scenery* (1946) and, characteristically, returned with *Man and the Land* in 1955.

It was a measure of his prestige that Fraser Darling was invited to join this august company, with the result that *Natural History in the Highlands and Islands* was published in 1947. Coming on top of his other books on the north-west of Scotland this firmly established him as a popular and influential writer on natural history, as well as on the economic and social problems of sparsely populated areas. *Natural History in the Highlands and Islands* was a great success in every sense, and its influence can still be traced in current books about the Highlands. However, Fraser Darling suffered from the British academic snobbery of the time and *Natural History in the Highlands and Islands* fell foul of the critics, particularly one, in some respects.

In *The Scottish Naturalist* there appeared over 2½ pages of almost vituperative nit-picking, some of it undoubtedly accurate. Then, as if he had recovered his balance, the reviewer changed key to become laudatory.

Clearly a book like this is exceptionally difficult to write and most of us would not have the courage to attempt it. Fraser Darling's views on conservation I most heartily endorse; his passionate love of his chosen land, and ability to inspire it in others, I admire and respect. This book has warmth and personality and an infectious appreciation of the good things of life.

The solution to this problem was for Fraser Darling to invite J. Morton Boyd (later Director of the Nature Conservancy Council in Scotland) to join with him in the revision of the book, which reappeared under joint authorship as, simply, *The Highlands and Islands* – a much more pretentious title than the original had had. (Morton Boyd had been inspired by *A Naturalist on Rona* to give up engineering and study biology.)

Not having read the review it was at this period I fell under the spell of Fraser Darling. His books were about real places and had the real adventure of man not taming, but learning to live in concert with the wilderness. He showed us what was wrong with a land we loved and also showed us that there were ways in which it could be changed for the better.

These were great years in the Highlands, with hopes of rejuvenation all around. Easter 1948 I spent with my Uncle Bob in Strathtummel, where he was one of the army covering the manmade desert of the hills for the Forestry Commission (at that time timber and timber products were the second largest item in British imports). All around were the massive engineering works of the Tummel-Garry hydroelectric scheme, bringing 'power to the glens'. In the summer we went cycling in the north-west, where we passed Isle Martin, where my grandfather's boat, the *Chrissie,* had been sold to a cutlery manufacturer as a service vessel for the flour mill he had set up on the island. Just along the coast was Tanera Mor, the island where Fraser Darling had evolved some of his ideas which seemed to be a basis for the transformation of crofting in the Highlands. At opposite ends of the Great Glen a Canadian distiller and the descendant of the Lord Lovat executed for his duplicity in 1745 were trying out cattle ranching on a big scale. Later there were to appear, again near both ends of the Great Glen, a pulp and paper mill and an aluminium smelter, moving us a little more in the direction of a Norwegian-type economy.

INTRODUCTION

Like many celebrities of our own time, Fraser Darling – or his publisher – thought that a children's book would now be a useful venture. The result was *Sandy the Red Deer* (Oxford University Press, 1949) illustrated by Kiddell-Monroe. The story is centred around Murdo Mackenzie the stalker and his children Roderick and Flora. The illustrations picture the 'Fraser Darling Country' dramatically and Loch Sheallag, Ben Dearg and Ben Fada are mentioned – although the latter two names are common enough in Highland Scotland.

That the book was strongly didactic can be deduced from the chapter titles:

Where Sandy and Sheena live
How Old Morag keeps watch
Sandy leaves his Mother's Herd
Sheena and her New Calf
What Sheena did when the Fox came to her Calf
How the Deer spend their Summer
What the Deer do when the Snow comes.

FIG 1
Roderick and Flora watch the deer
(*Sandy the Red Deer*, Oxford University Press)

How does a writer finish off a narrative that is essentially a record of a cyclical process? Winter comes and with it the snow. The deer move out of the glen towards those bare patches of heather and sedge where they can feed. Fraser Darling emphasises the matriarchal structure of the herd by saying that: 'Old Morag was at the head of her string and Sheena was the last'. (The last mention of Sandy was eight pages earlier). The last sentence is given to the children's father, the stalker: 'They'll come back again soon' said Murdo, 'come with me now and help me build a snowman'. A Highland '*il faut cultiver notre jardin*!'

There is no evidence to suggest that there was any danger of Fraser Darling ousting A. A. Milne or Kenneth Grahame as a children's author.

Simultaneously, he was working on the West Highland Survey. Tom Johnson, the great reforming Secretary of State for Scotland, set up the West Highland Survey in 1945 and commissioned Fraser Darling to direct it. Who better, with his course record? The survey itself was carefully designed, executed and analysed. It covered 'A Brief Historical Résumé of the Highland Problem', background –

FIG 2
The herd in winter
(*Sandy the Red Deer*, Oxford University Press)

relief, land forms, vegetation and communications, population, the ecology of land use, the agricultural situation (including crofting), the social situation and a summary.

West Highland Survey: A Study in Human Ecology (Oxford University Press) did not appear until 1955. Behind the scenes there was a degree of hostility from vested interests. Fraser Darling was probably naïve politically in expecting competing interests to cooperate and make sacrifices for the common good. He also saw the need for substantial resourcing. Like an Old Testament prophet he was 'filled with wrath at our wilful blindness and the indifference we show to the wasting of our inheritance' and was reluctant to make emollient changes to his text.

A poor reception by the politicians and the opposition by the lairds to the publication of deer numbers on their estates increased his disillusion till he felt compelled to concentrate his efforts furth of Scotland.

About this time, with my mother, I went to a lecture of his to the Royal Scottish Geographical Society. Fraser Darling was quite clearly a superb communicator in print, but I have to say that I was not impressed by his stage presence and delivery. In retrospect I was probably expecting a rerun of his books about reclaiming the land on Tanera Mor or life among the seals and the deer. By this time, however, Fraser Darling had moved on and would have been thinking aloud with us about the growing environmental crisis.

I have recently listened to some tapes of Fraser Darling in full flow in the Reith Lectures of 1969 and in studio discussion. The lectures were obviously meticulously prepared. His reading of the text was excellent, clear and nicely timed. His message was complex and fairly unfamiliar at that time. 'Population, pollution and the planet's generosity' might then be unfamiliar as interrelated concepts but were well explained in straightforward language. Refreshingly, there was a twist of wry humour and the occasional Darlingesque dismissal of pettifogging detail.

The studio discussion was obviously unscripted and in it Fraser Darling demonstrated his skill in thinking on his feet. When challenged on the question of his personal contribution to world over-population he was able to point out that, while he had been responsible for four children, he had had three wives (i.e. only one child per adult), had taken the view that he should have no more children and felt no guilt about his situation.

I have to accept that my oral memory is probably unreliable and that my earlier disappointment was an indicator of a certain amount of immaturity.

Fraser Darling by this time was probably appreciated more abroad than at home. He felt it necessary to extend his activities to the United States and East Africa, from 1959 to 1972 enjoying a roving commission researching, publishing and proselytising. At a time when 'the wind of change' was blowing through Africa he was able to produce sound planning material, but it is doubtful if much concerted action stemmed from it. His main significance lay in the part he played in setting up the National Parks in East Africa.

From this period emerged his advocacy of niche diversity, the deliberate conservation of native species of plants and animals instead of the mass introduction of cattle, sheep and goats. His criticism of over-burning was clearly an extension of his criticism of over-burning in the Highlands. 'A good friend but a bad master' he might have said. Again, his Highland experiences shone through when he emphasised the importance of reconnaissance on foot. We are all familiar with sequences showing vehicles racing across the African plains hunting down terrified animals. For Fraser Darling aeroplane, helicopter, jeep or canoe were acceptable in getting to an area, but it was important then to walk on foot.

His American reputation took off after 1956 when *Pelican in the Wilderness: A Naturalist's Odyssey in North America* was published. On the face of it a survey of the US National Parks, it turned out to be a critical analysis of much that was wrong, not just in the United States but in most of the developed world. That he had by now learnt something of diplomacy is shown by a reviewer's criticism that 'Darling seems too polite in avoiding sharp criticisms of practice and research, of federal and state agencies concerned with conservation.' Yet it was a seminal work.

He returned to this country to give the Reith Lectures (*Wilderness and Plenty*), important in three respects. Someone at the highest level had now recognised that Fraser Darling had something worthwhile to say – his views were now to be accepted as orthodoxy. Fraser Darling was not alone, in 1969, in having concern for the way in which the world was being mismanaged by man, but his unique contribution was to knit together the different components leading towards the environmental crisis and to explain these in simple language to a wide audience. Reference to the simple time-line below will show the crucial importance of the Reith Lectures in the growth of environmentalism.

For Fraser Darling the lectures were a personal turning point. They caught the attention of Harold Wilson, the then Prime Minister, and Anthony Crosland, who became the first Secretary of State for the Environment in a pre-election reshuffle the following summer. Crosland set up a Royal Commission on the Environment. Fraser Darling was appointed as one of its founder members and remained on it until his final retirement from public business in 1973. One of Harold Wilson's last acts in the week he lost the 1970 election was to propose Fraser Darling to the Queen for a Knighthood. This gave Fraser Darling a modicum of amusement – it was obviously what did for the Labour Government!

Max Nicholson (1904–2003), himself an environmental guru of the first order, who forfeited the highest rewards as a result of upsetting the establishment, summed up Fraser Darling thus:

> I knew him for 30 years, and had sometimes to share his sufferings over actual or imagined setbacks, but in the end he came into his own, and found a receptive audience for a contribution that was partly scientific, partly ethical or philosophical, and at times even mystical. He belongs at the far end of a spectrum that extends all the way from the most practical or political of conservationists to the poets and dreamers. Both extremes are needed even if the task of getting them to mix can be demanding.

In *A Herd of Red Deer* we see the young Fraser Darling presenting thoughts which have already gone beyond straightforward objective observation to wider considerations that were to make colleagues uncomfortable throughout his life. In later life he was to reminisce that he was told by an eminent academic that he would never be appointed to a Chair if he pursued this kind of research!

A Herd of Red Deer

IT IS DIFFICULT to over-estimate the importance of the publication, in 1937, of *A Herd of Red Deer*. With it, Frank Fraser Darling marked the end of his apprenticeship as an ecologist. Up till then his published work showed the specialisation of the young scientist and the results of small studies, each with a definable beginning, middle and end. There were papers on:

> *Efficiency in Dairy Cattle, the Problems of a Criterion* (1932),
> *Studies in the Biology of the Fleece of the Scottish Mountain Blackface Breed of Sheep* (1932),
> *The physiological and genetical aspects of sterility in Domestic Animals** (1932),
> *A Note on the Inheritance of the Brindle Character in the Coloration of Irish Wolfhounds** (1933),
> *Mendelian Inheritance and the Chromosome Theory* (1933),
> *Animal Husbandry in the British Empire* (1934).

Those marked * were co-authored.

With *Animal Breeding in the British Empire: A Survey of Research and Experiment* (1937), it was clear that he was now ready to move on to bigger topics and a higher level of generalisation.

As we have seen, *A Herd of Red Deer* took the reader out of the library and the laboratory and into the reality of the tough climate and landscape of Wester Ross. Today we are beset with television nature programmes which frighten and threaten or invite us to admire the courage of the commentators. In *A Herd of Red Deer* there is no whispered voice-over: 'This is potentially very dangerous.' Only by inference can we realise what an achievement it was to keep an almost constant watch on these highly mobile animals in one of the toughest environments in the British Isles.

On certain days of the year Fraser Darling felt obliged to make special journeys to map as many deer as possible at one time. He used the high ridges to save time and distance. 'Such days mean 35–40 miles of walking and 7,000–10,000 feet of climbing, and they are among my most pleasant memories'. In this rugged terrain he reckons that one mile per hour represents good progress and ruefully notes that the deer move easily at six times that speed.

A personal note. Quite recently I noticed that the only considerable hill in Wester Ross I had not climbed was Beinn a'Chaisgein Beag (2,232 feet – The small hill of cheese). So it had to be climbed and there was only one particular July day when it could be done. In his chapter on 'The Country' Fraser Darling has a very interesting paragraph in which he describes the head of Uisge Toll a'Mhadaidh as a 'strange place' with 'sensations'. This had to be tested and could easily be added on to the hillclimb.

At lunch-time on the previous day it began to rain and it continued to rain for at least 48 hours. The walk began at sea level at Gruinard Bay and the first stretch was up a steep-sided valley – 'almost Himalayan', to quote the guidebook. Two superb waterfalls, the water creamy-brown with peat, added a touch of drama to the sense of remoteness.

The next stretch was over and through a mass of low hills, wet, depressing and difficult to navigate through. One of the things Fraser Darling is good at is finding beauty and interest in seemingly ordinary landscapes. For most hillwalkers Lewisian gneiss is the tough, ancient, shapeless and frustrating basement for the spectacular giants of Liathach (Torridonian sandstone) and Beinn Eighe (Cambrian quartzite). But for Fraser Darling 'the country as a whole has a joyful quality, and the constant changes of lights and shades ... are stimulating to the seeing eye.' A straightforward description of vegetation and animal life concludes: 'Many and happy are the nights I have spent on this ground. How often have I been wakened at dawn in May and June by the piping of the ring ouzel, the harbinger of each new day!'

Fraser Darling was well aware that his departures from purely objective description might raise some hackles, but he was unrepentant, maintaining that subjective factors did have an effect on the perceptions – and hence the behaviour – of man and other animals.

My next problem was to cross the Uisge Toll a'Mhadaidh, the big burn draining Loch Toll a'Mhadaidh in the big corrie – the 'strange place' – where 'there is discomfort sufficient to make a man move' and the 'sensations' were to be experienced. The rain had made the obvious crossing point just impossible. A little upstream was a stretch with plenty of boulders where one could skip across. Unfortunately, there was no far bank – only a sheer rock-face over 100 feet high. A trudge upstream of about two miles brought me to a place where three large boulders got me across and I set myself to a long wet ascent of my target mountain on a compass course over 'fallen rocks, deep peat hags, and heaped moraines' (Fraser Darling's words).

One objective attained, it was clearly impossible for me to achieve in this watery wilderness the other, to relax and empty my mind so that I could pick up sensations 'caused through the eye by the dispositions of masses and planes and their relation to the course of the sun'. So I turned and retreated as fast as I could.

Was my little expedition a total failure? Not at all. My admiration for Fraser Darling now had a solid basis. I had had a tough experience in July. This man had covered the same ground in all weathers, at all seasons, not just walking but waiting and watching and writing things down.

My first draft contained a commentary on Fraser Darling's use of Imperial units – no longer necessary since it now seems that we are to turn the clock back and further confuse young people trying to fit into the 21st century. No doubt technical colleges will soon be measuring their windmills in Danish thumbs and holding drystane dyking courses in megalithic yards.

Fraser Darling's approach is more qualitative than quantitative but he does use figures – usually for comparison, rather than as absolute values. Thus we have TABLE 1 on p34 listing deer densities from one to 12.5 acres to one to 300, with a note on p45 stating 'the figures for reindeer in Alaska are not sound.' In FIG 8, showing temperature and humidity for three days in October 1935, it is the changes in atmospheric conditions which are related to observed changes in the behaviour of the deer and described in pages 111–112.

In 1937 temperatures were measured in standard conditions in a Stevenson screen with the thermometer bulbs at four feet from the ground. However, Fraser Darling wanted a 'deer's-muzzle' measure of temperature and humidity and therefore made his own recordings at two feet above the ground.

The One-Inch Ordnance Survey map Fraser Darling used showed the bench mark at the Dundonnell Hotel accurate to 0.12 inches above sea level. On MAP 6, in the south-east corner, can be seen Beinn A'Chlaidheimh. The OS map then showed this as a long, steep-sided ridge with a rounded top, the summit contour being 2,750 feet. This fine mountain reached 2,960 feet in 1953 (Corbett's Tables). In the First Metric Edition of Munro's Tables (1974) it had grown to '3000c feet' on the six-inch map and 914 metres on the Metric 1/50000 map (3,000 feet = 914.4 metres).

Of this area and the mountain's shape Hamish Brown talks of 'a whole muddle of Munros, which even the OS failed to delineate until 1974. The compass is not always reliable and the map not always accurate.' On Beinn A'Chlaidheimh itself 'you felt and sensed crags and slopes rather than saw them.' 'The summit ridge is quite narrow' says the *Scottish Mountaineering Club Hillwalkers' Guide*, laconically. The editor's own recollection is of a misty morning climbing over big Torridonian sandstone blocks like a rather badly-built staircase leading to the summit ridge, unproblematic technically but with an uncomfortable feeling of great depths on either side.

The book's first chapter sets the scene for the later description of the research. 'The Country' describes the research area, its physique, its geology and the plant and animal life associated with its subdivisions. Right from the start he uses subjective language to enhance his objective descriptions – a February day is 'cheerless', 'white heads of bog cotton bob cheerily in the wind of a June day'. In Strath na Sheallag he asks: 'What are the reasons for the charm some places have for beasts and men? Shelter and a fresh bite' – in other words, the measurable factors – 'are not a sufficient explanation.' The weather receives full attention and the chapter closes with a justification for departing from objectivity:

> The country as a whole has a joyful quality ... I have not fought the country these two years but have let it be my foster-mother. Her discipline has been stern but her smile is never far away.

The chapter on 'Technique and Personal Relations' reads like ancient history. He gives precise details of suitable clothing – 'Khaki handkerchiefs were chosen in preference to white ones, and plenty of them carried, for the nose runs more than normally when moving about on the hill.' 'Waterproofs should be avoided completely.' 'Harris tweed is almost silent.'

The book was illustrated with excellent photographs in black and white, Fraser Darling's own, which are reproduced in this new edition. His comments are interesting. In 1937 most outdoor photography would have been done with heavy equipment, but: 'You cannot carry a large camera many miles on the hill.' Instead he used a Leica with automatic focussing and a range of lenses – state of the art for the period. Even so, 'Photography of deer on Highland hills has been a great disappointment. The animals blend so perfectly into the background that the monochrome print shows no contrast.' (Even with modern colour film the fact that 10 per cent of males have a colour perception problem means that the red deer on a dun hillside is very difficult to spot.)

Despite his reservations, most of his subjects are more than mere illustration. The photographs on p 187 are an obvious pair of shots and, under the caption to the lower one, Fraser Darling added: 'Note the position of the fore-feet in these pictures. At gaze, the fore-feet are always brought together, giving the head all possible height.' So obvious, but so cleverly demonstrated.

Perceptive readers will have noted that the contemporary colour images are more than attractive pictures. The scene having been set in the first two we follow the red deer from one winter to the next, with some of the subjects closely matching Fraser Darling's originals.

His views on stalking and techniques for the raising of awareness are of interest and clearly based on many hours of experience.

'Territory and Population' are spread over two chapters and are covered by reference to earlier authorities, then comparing them with his own observations, which are very detailed and occasionally stray beyond sterile objectivity – 'It is amusing to watch a hind having a rub.' Three kinds of territory are described and plans in the text introduce a note of precision. Overcrowding and its antisocial effects are noticed – one example of where Fraser Darling made the initial discovery, with a straightforward explanation, for a later generation to come along with something more sophisticated. The danger of extinction and checks on population growth are interestingly covered – not least because this anticipates his work in the 1950s which resulted in *Pelican in the Wilderness*.

'Territory and the Social System' takes existing authority and builds on a mass of field evidence, with a side-swipe at *Monarch of the Glen* – painted by 'a faulty observer'. 'Some Social Factors and Comparisons' has a deal of interesting observation on The Voice, Play (where Fraser Darling has interesting speculation on why young animals play – if that is the right word), and Sociality in general.

Two chapters follow on 'Movement', one on 'The Influence of the Weather' and another on 'Insects and Food Supply'. There is an excellent diagram which shows how the various factors interact to affect the deer's behaviour. This is followed by a wealth of detailed observation and a recognition of emotion in the stalker who refuses to shoot a hind when bad weather is about to force the deer down to low ground.

The passages on insects make for horrific reading. The cleg, whose life-history had only been worked out and described three years earlier, gets several pages. One has a mental picture of Fraser Darling lying low in the heather, filling in time by recording times, dates and frequency of cleg bites. *Culicoides* – the Highland midge – is covered as a pest of the deer, but, with admirable self-restraint, Fraser Darling says nothing about its effect on him. Those who know the Highland midge will know how he must have suffered agonies from visitations when the wind dropped, the temperature rose or the air was the least touch damp.

The chapter on 'Reproduction' was:

> at once a most interesting and disappointing chapter to write, because of the variety of the problems raised and left unsolved.

A big chapter, there is a great deal of interesting observation, a well-known diagram linking the yearly cycle and sexual growth, and even diary extracts. At one point he marvels that the tips of the antlers may be used to masturbate several times a day and yet:

> These antlers, used now so delicately, may within a few minutes be used with all the body's force behind them to clash with the antlers of another stag ... at one moment exquisitely sensitive, they can be apparently without feeling the next.

On the subject of the sensitivity of tips of antlers, the editor remembers a late afternoon in Loch Ossian Youth Hostel when a magnificent stag came in, scattering excited Germans in all directions, and threaded his way past lines of washing and a red-hot stove chimney to get his daily supplement of Vitamins C and E, prepared for him by the warden.

The chapter on 'The Senses' concludes:

> This short account of the senses of red deer is quite inadequate and shows the limitations of observation.

Perhaps. The reader can judge for himself.

Fraser Darling's 'Conclusion' begins:

> This book is not an attempt to propound a theory of animal behaviour, for as yet I do not have one.

This statement is best left to the next section, when we examine the relevance of *A Herd of Red Deer* today.

INTRODUCTION

Relevance Today

IT WOULD BE naïve to suppose that all that Fraser Darling wrote in 1937 would be acceptable in the 21st century. If we think of the enormous changes and discoveries since then – the nuclear bomb and its successors, the jet engine, the space race, the explosion of scientific knowledge and research, the communications revolution – it must be obvious that there must have been much work carried out in his field also. Fraser Darling's methods and conclusions might seem to be outmoded, even quaint, and irrelevant to our times, a modern equivalent of Sir John de Mandeville's *Travels*.

Frank Fraser Darling's position in 1937 was similar to that of a geologist at the same period. Over a century of field and laboratory work had amassed a huge amount of detailed knowledge. The composition of rocks, their age and order of formation was known. How they were formed and the sum of small movements that had subsequently affected them were known or could be worked out. For example, Peach and Horne spent some 20 summers unravelling the mysteries of the north-west Highlands, so that the Moine Thrust and surrounding territory has become a classic to be studied by each generation. As the great Austrian geologist Suess said, Peach and Horne 'have rendered our northern mountains transparent.' It was even known that huge movements on a continental scale must have taken place. But the really big question was not answered.

I have a copy of Arthur Holmes' *Principles of Physical Geology* of 1945 ('t'best book on t'soobject') which ends with a discussion of Wegener's Hypothesis of Continental Drift of 1910. Note – hypothesis. The very last section is 'The Search for a Mechanism', with two very interesting diagrams: 'to illustrate a purely hypothetical mechanism for 'engineering' continental drift.' Now, as we know, every 12-year-old can talk glibly about plates and subduction and the rest.

The reader will recall that Fraser Darling's stance was that the great bulk of papers on animal behaviour in his time lifted the organism from its normal environment and placed it in a set of artificial conditions. For a variety of reasons – not least the stress of being observed in an artificial environment – the results were not valid for interpretation of representative behaviour. As described in the book, preliminary studies of animals in their natural surroundings should be the initial steps to any experimental approach. Such studies were of rare occurrence until after he had shown the way.

The straightforward account of what he observed and how he organised the operation is still well worth reading today. Some of his extended descriptions, like his observations of young deer at play, or the rutting season, or the behaviour of deer in bad weather are absorbing and – because they are a record of his observations – are true.

We know that every piece of worthwhile research throws up as many new questions as it resolves the original ones and this is very much the case with *A Herd of Red Deer*. With the passage of time other workers have tested his hypotheses and refined his findings in the light of new methods and new knowledge, to find his observation and recording reliable.

For example, in the chapter on 'Population' he discusses overcrowding, estate policies and the danger of extinction. When we turn to the 1994 Scottish Natural Heritage Policy Paper 'Red Deer and the Natural Heritage' we find a series of graphs illustrating the effects of increasing density of hinds on calving rates (decline), increasing density and juvenile mortality (mortality increases, males faster than females), increasing density and antler weight (antler weight falls), increasing density and fertility (the percentage of yearlings falling pregnant drops to zero, but more slowly in the forest than on the open hill). Clearly, these results are the tip of an iceberg of many hours of observation, calculation and measurement in conditions that would have been familiar to Fraser Darling, but they do not invalidate his initial work to any degree. He thought that extinction would happen as the environment was degraded and the factors above came into play at the same time. Current thought sees extinction as a two-phase process; first the degradation of the environment, followed by a population crash.

The reviewer who filleted *Natural History in the Highlands and Islands* had to concede that: 'This book has warmth and personality and an infectious appreciation of the good things of life'. The same might have been said of *A Herd of Red Deer*. Fraser Darling was well aware of the risks he was taking with his approach. As we have already seen, he was happy to show enthusiasm and to depart from objectivity when it suited him, believing as he did that there was more than just counting and measuring and that psychological factors were important.

The last few words are interesting.

> ... amongst the higher animals we find response to sets of conditions and a spontaneity of action which we, as so-called rational beings, could not better. In some instances I feel that the most simple explanation of an act of behaviour is to follow the bare outline of our own mental processes in such a situation ... Who are the people with whom the higher animals are most serene. And who achieve most success in their management and training? Not those who look upon them as automata, but those who treat them as likeable children of our own kind.

We may not agree scientifically with what he says, but it is interesting and attractive and shows how influential Fraser Darling could be in encouraging empathy between man and the natural world.

Nowhere is the march of time more evident than in his 'Conclusion'. Earlier it was noted that the book was not an attempt to propound a theory of animal behaviour, for as yet he did not have one. He believed the teleological approach to animal behaviour, with its insistence on some higher purpose, to be dangerous, but he also believed that the objection to anthropomorphism current in his time could be overdone. Other ideas are set aside. It may be that the quotation above is being trailed tentatively past us.

What is almost self-evident is that work on all aspects of the red deer broadened and deepened. Fraser Darling himself was appointed by the Nature Conservancy in 1952 to survey the Scottish deer population and the first counts started in 1953. The Deer (Scotland) Act of 1959 provided the legislative framework for the management of wild red deer in Scotland. The island of Rum became, in effect, a large open-air laboratory for the study of the red deer. Here it was possible to study the deer as Fraser Darling did, on their own territories, but with fewer distractions. There was the advantage of continuity of research: instead of a 'snapshot' of the deer communities they could be studied through the generations. Above all, the fieldwork could go hand-in-hand with work in the laboratory, and understanding could develop in parallel with understanding gained from the wider scientific community.

For example, in his section on 'The Course of the Rut', Fraser Darling describes the behaviour of 'Wide-spread', a big dark-coloured stag. A group of 77 hinds and followers occupy about 20 acres and for about two days are completely dominated by the big stag, who chases away any rivals. However, some young stags and a few rutting stags hang around and, as time passes, these stags begin to encroach and detach 'harems' of their own. After nine days, 'Wide-spread' disappears and there are 10 harems occupying the area, with 12 more rutting stags in the vicinity.

This behaviour is best explained in terms of 'the selfish gene', a neat term first introduced in 1976. Two strategies for the continuation of the species are being demonstrated here. There is the straightforward – what, in my unscientific way, I would call 'hash and bash' – dominance of the strong way. But there is also the cunning, biding the time way, sniffing round the edges, seeking a weak spot to cut in and drive out a suitable hind. The modern research has established that there are two kinds of stag here, with slight genetic differences that affect their behaviour and will be transferred to their progeny.

When *A Herd of Red Deer* was being researched and written up by Fraser Darling this kind of knowledge and understanding was not available to him, so that it has to be said that his interpretation – as opposed to his observation – must be regarded as doubtful today.

A present-day examiner might make a report like this on *A Herd of Red Deer*:

> Observation and Description – excellent,
> Empathy and Imagination – excellent,
> Explanation – requires further work.

INTRODUCTION

Fraser Darling and the Environmental Movement

1937	*A Herd of Red Deer: A Study in Animal Behaviour*
1938	*Wild Country: A Highland Naturalist's Notes and Pictures*
1939	*A Naturalist on Rona. Essays of a Biologist in Isolation*
1940	*Island Years*
1943	*Island Farm*
1947	*Natural History in the Highlands and Islands*
1949	*Sandy the Red Deer*
1955	*West Highland Survey. An Essay on Human Ecology*
1956	*Pelican in the Wilderness: A Naturalist's Odyssey in North America*
1960	*Wild Life in an African Territory*
1962	*Silent Spring* – Rachel Carson
1963	Red Deer Commission established
1969	*Wilderness and Plenty* – The Reith Lectures
1970	*Royal Commission on the Environment*
1972	*Blueprint for Survival* (The Ecologist)
1972	*The Limits to Growth* (The Club of Rome)
1974	*Small is Beautiful: Economics as if People Mattered* – E. F. Schumacher
1992	Rio Earth Summit and Local Agenda 21
1996	Deer (Scotland) Act 1996
	Deer Commission for Scotland established

CHAPTER ONE

The Country

THE DEER FORESTS of Dundonnell, Gruinard, and Letterewe in Wester Ross form the north-western corner of the largest continuous tract of forest ground in Scotland, that which stretches northward from the foot of the Great Glen. The parts of these forests which form the subject of this chapter cover an area of about 80 square miles, or 52,000 acres, of this mountainous country. This area is uninhabited except on its northern edge, and there the people are few.

The central mountainous mass of An Teallach dominates the countryside, and this range of Torridonian Sandstone is considered the wildest on the Scottish mainland. Glaciers of an earlier age have gouged out deep corries, scored the hillsides in slanting fashion, and left moraines of worn boulders below the corries' lips. These Torridonian peaks rise sharply from almost sea-level to over 3,000 feet on the western side and then fall precipitously to their corries at 1,750 feet on the east. 'S Bidein a' Ghlas Thuill reaches 3,483 feet at the head of the Glas Thuill corrie. Sgurr Fiona (rightly Sgurr Fheoin), 3,474 feet, falls almost sheer, and with the precipices of Sail Liath, 3,150 feet, bounds the Toll Lochan corrie. Sail Liath and two other outliers of the group, Glas Mheall Mor and Glas Mheall Beag, are capped with quartzite boulders. A sparse vegetation of alpine poagrass, fine-leaved sheep's fescue, alpine tufted hair-grass, stonecrops, and *Rhacomitrium* moss clings on the cyclopean terraces of the Torridonian Sandstone, but on the quartzite caps there are acres without a plant. *Rhacomitrium* moss grows in patches here and there where the stones lie fist-size for a space. Yet it is on these high barren slopes that the few snow buntings stay for the summer and nest among the larger boulders. The ptarmigan is about the tops also, preferring the gravelly sandstone to the shiny quartzite, but its numbers are kept low by the golden eagle, which hunts there each day.

The corries immediately below the Torridonian peaks are as dry as may be in a rain-washed place and free of peat. The alpine herbage is sweeter than that of the peat, and the mountain juniper grows tortuously and in poverty, flat to the gravel. (That adjective of common usage – 'sweet' – applied to herbage signifies a fair content of mineral salts, desirable in the food of herbivorous animals. 'Sour' herbage, such as grows on waterlogged peat, is minerally deficient, and consequently low in proteins.) In the Glas Thuill springs of water rise from the hill face, and their track is green till they join in a sandy-floored burn of crystal clearness, unsullied as yet by the acid of the peat. The grazing faces of the Glas Thuill and Toll Lochan corries are the summer haunts of deer and a few feral goats, but in winter, when the snow lies above the 1,700-feet contour and sometimes below it, the corries seem empty of animal life. But they are not empty, for as I stand on the shores of the deep Toll Lochan below the buttress cliffs of Sail Liath on a cheerless February day of scudding mist and wet snow, a skirling flock of fieldfares may pass overhead, and on the sand at the lochan's edge a dipper sings his small sweet song – and the day is changed.

The wind plays queer tricks in these corries, especially in that of the Toll Lochan. If a south or south-west wind is blowing hard, the waves on the Toll Lochan will be beating with considerable force against the buttress cliffs at the western end, and the observer, looking up, sees the clouds flying fast in almost the opposite direction. Waterspouts disport themselves across the shallow eastern end of the lochan. This whipping-in of the wind into these corries is in accordance with aerodynamical expectations, but on such a scale it is remarkable. When an east or north wind blows into the corries, it finds no gentle slope to deflect it upwards. The cliffs receive the full blast and there is again a reversal of direction and a turbulence of the currents at the head of the corrie. It is difficult and almost impossible to stalk deer there if the wind is strong. A snowstorm demonstrates the air movements faithfully and in the Toll Lochan corrie provides a very fine spectacle. These remarks apply in a slightly lesser degree to the eastern corrie of Beinn Dearg Mhor.

The waters of the Toll Lochan and Glas Thuill corries fall into

Coir' a' Ghiubhsachain and join to form the Garbh Allt, a boisterous torrent of many deep falls which runs into the Dundonnell River. Coir' a' Ghiubhsachain is a desolate place, sandstone-shelved on the Teallach side and bounded to the south-east by quartzite cliffs up to 250 feet in height. The floor of the corrie is rockstrewn, and the shallow wet peat bogs along the course of the burn hold the trunks and roots of primeval pine-trees. These, probably, are the origin of its name rather than living trees, of which the place is devoid except for a dwarf rowan growing from the fissures here and there. Deer do not like the continually wet, waterlogged wastes of Coir' a' Ghiubhsachain, yet the corrie is a link between winter and summer grazings, and they make for its difficult ground when disturbed on Carn na Carnach. Grass is scarce in this corrie, for apart from the sheer poorness of the ground, half the area or more is bare rock. Banks of heather grow well at the foot of the quartzite cliffs, but for far on each side of the burns the vegetation is of bog cotton and other sedges, some sphagnum moss, lichens, and bog asphodel. The wet bogs of all this country are starred golden with the asphodel in late June and July, and in September and early October its red head of seedpods brightens the cloak of withering sedge. Later they bleach, but endure as hard-bitten relics of summer, standing stiff and white on a winter's day.

Stretching southwards from the head of Coir' a' Ghiubhsachain is an area of quartzite slabs, and the same formation falls eastwards into Gleann Chaorachain from the corrie's cliffs. Quartzite slabs, which are not loose boulders but bedded rock, give a poor hold to herbage, especially if they are well slanted. The rock is hard and shiny and no foot is safe on it. Weather, peat acid, and plant roots do not disintegrate it appreciably, and the thin covering of wet peat sloughs away in patches where there is no obstacle. The quartzite ground is always wet because the rock admits no moisture, and perhaps for this reason the palmated newt is common in the pools dammed by the peat on the shelving rock. Near the southern boundary of the Gleann Chaorachain quartzite is a sharp fault giving a sheer cliff of from 30 to 80 feet, running over into Coir' a' Ghiubhsachain as a gully and forming a narrow pass. The fault

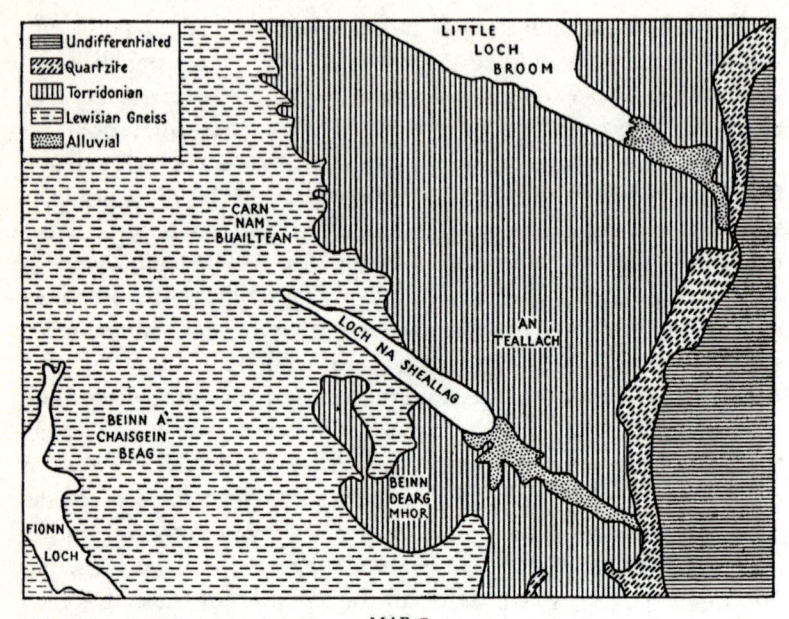

MAP I
Geological sketch-map of the terrain.

is pierced by a deer path, as all such places are where there is a break giving foothold, but the cliff fashions a territorial boundary for two groups of deer, and for this reason I shall have cause to mention it again. The herbage of the quartzite slabs is poor in itself, and this geological formation is inhospitable to man and beast wherever it appears. Red deer hinds will not make their home on it, and on its passage through this country it holds no human habitation.

Carn na Carnach is a different place, of green banks and hollows and ample shelter from all winds. It is the only part of my beat composed of the undifferentiated eastern schist – a formation closely stratified and containing more alumina than the neighbouring rocks. It is more easily crumbled into soil by the constant action of weather and growing plants. The Carn is a cheerful place of grass, birch and alder trees, and good heather. The sward is made bright in spring with primroses and milkwort, and later with tormentil and heath bedstraw. Rabbits are plentiful, in spring and summer it is beloved of wheatears, ring ouzels, and cuckoos; and the robin and wren, the rock and meadow pipits live there the year round. On the north-eastern slope, called Achachie, 'the field of little fields', are the grass-covered foundations of summer shielings, dwellings disused beyond the memory of living men. Carn na Carnach is the home of hinds, a place rarely free of deer for long. The burn which bounds the north-western edge of the Carn is also the meeting-place of the quartzite and eastern schist formations. The division is vividly sharp when spring breaks on the country – a chequer-board of light-grey rock and brown herbage on the one hand, a steep green sward with its teeming life on the other.

Gleann Chaorachain and the strath of Dundonnell are unusually well wooded, primitive birch scrub and alder for the most part, with a few sallow, gean, and crab-apple trees. There are large numbers of planted trees round the cultivated land of the strath – oak, ash, beech, chestnut, lime, larch, and pine – and furze and bog myrtle are thick among the moorland herbage. In June the ground is scattered with rosettes of sundew and butterwort, and the air is bright with dragon-flies of many colours. The fauna of the glen is varied and numerous. Buzzards and kestrels live in the wooded cliffs unmolested, and the wild cat has a stronghold among the fallen rocks. The red squirrel is common and makes his rounds in season of the trees' fruits. The squirrels are more tame here than I have found them elsewhere and a source of pleasure to the few of us who live in the strath. Rabbits are too plentiful, and there are hedgehogs, moles, field and bank voles, and long-tailed field-mice, all busy in their season among the leaf-mould, moss, and tree-roots. Roe deer are in the woods – shy, elfin creatures – and each year otters try to establish themselves on the river. In spring there is a migrant surge of willow-wrens and redstarts, and flocks of tits, fieldfares, redwings, and blackbirds come for the winter. There are no adders here, but plenty of common lizards. This strath is the refuge of many deer in times of heavy snow. The nearness of the western sea makes this part of the Highlands a paradoxical country, for on a New Year's Day when the hills were bare and An Teallach a white, ice-bound peak, I have found a primrose blooming in the shelter of the rocks and dead fern by my house in the glen, and all through the hard winter of 1935–6 a few plants of herb Robert bloomed continuously among these same rocks. By the end of March 1935 the frogs were croaking happily in many a peaty dub, as high even as the floor of the Glas Thuill. I pitied them, for at that time there should be no spring song in the Highlands. My fears were justified in May, when frost and snow descended on the country. The thaw came soon and there were thousands of dead frogs in the pools. By the outstretched position of the arms it was apparent that most of them were females. Only the few had their hands clasped. It was remarkable how soon their carcasses disappeared. On one occasion I saw a stag wading in a shallow dub and eating them. The stag, craving minerals when growing his new antlers, becomes quite resourceful in trying to satisfy his appetite for lime, and ready to become carnivorous for the moment.

There are several hundred acres of pine wood on the south side of the glen, and the last few trees reach up to the 1,000-foot contour. The trees were planted nearly 100 years ago, but the wet troughs of peat have killed many of them and only those on the drier knolls and braes have thrived. The wood is a rough stretch of clumps of pines reminding me of those few areas of the ancient Caledonian Forest which still survive, but the pines are peaked and not round- or flat-topped like the old trees of the country; and there is no wealth of juniper and blaeberry below them as in Rothiemurchus. The wood holds its group of red deer and a few roe deer as well. The trees are constantly worked over by coal-tits and gold-crested wrens, and in 1934 I saw a pair of crested tits here, immigrants to the west perhaps from the Moray Firth country, over the bleak pass of the Dirie Mor. The slopes of An Teallach are steep and extremely rough between the pine wood and the head of Little Loch Broom and below Coir' a' Mhuillin. Terraces and cliffs of the Torridonian crop up irregularly, and the burn from the corrie falls deep through a cleft like most of its kind on these sandstone faces. These small and noisy waters attract the grey wagtail in summer, and the pale golden flash of the active birds adds to the vivid quality of the northern summer.

Northwards from An Teallach, beyond Coir' a' Mhuillin, lies a

Sgurr Fiona and the Toll Lochan Corrie
Coir' a' Ghiubhsachain in foreground

high plateau, 2,250–2,500 feet, free of peat and pleasantly undulating after the difficult ground of the mountain. Small flat stones lie edge to edge over almost all this curious place, natural crazy-paving in the garden of Boreas. Thrift, stonecrops, and lichens grow in the chinks, and there is no grass or moss except here and there where sand has collected in low dunes and shallow depressions. The plateau is waterless and dry in a dry time. The mountain hare is a species which fluctuates sharply in numbers, and during the time of my work here they have been at their lowest. I have seen only two of these animals in two years, each time on this high plateau of Meall Bhuidhe, where there is no cover and the barest of herbage.

The western slopes of Meall Bhuidhe and the rocky promontory of Sgurr Ruadh, the farthest outlier of the Teallach massif, form the sides of Coire Mor an Teallaich, an immense hollow, thinly clad by alps of *Rhacomitrium*, fine-leaved sheep's fescue, and typical alpine herbage. The upper reaches are free of peat, but below Lochan Ruadh and its red sandy shores is a great waste of peat hags and rubble from dead moraines. There are no trees, bushes, or brackens in this weather-ridden place until the lip of the corrie is passed, where the Allt Airdeasaidh falls roaring through its deep cleft in a succession of cascades to the sea. The grey fox of the Scottish Highlands and the wild cat live on Sgurr Ruadh and in Coire Mor, the eagles hunt over it regularly, and it is summer ground for the deer. On the shortest day of 1934, when the sun shone its full span, deer were grazing in multitudes on the corrie's sides and on the mossy stones about the tops. I saw a pair of dippers playing in the still water of Lochan Ruadh. They went through a dance of many curtsies on the sand and sang their little song. Coire Mor has a trace of summer kindness left in December, but in March and April the place is stark and snowbound.

Across the Allt Airdeasaidh is the triangular range of Sail Mhor, Ruigh Mheallan, and Sail Bheag. These hills form the boundary of the Torridonian Sandstone at the foot of their slopes on Lochan Gaineamhaich. The Coire Mor side carries a thin covering of poor heather, the more gentle north-western face is wet and covered with sedges and bents, but Sail Mhor itself is a very poor brashy hill, not frequented by the deer. Sail Mhor is an outstanding landmark, varying in its shapes when seen from different points of the compass. Looking towards the hill from the ridge of Sail Bheag it appears to be the upper half of an immense sphere, and the outline is regular enough to give the illusion of the sphere being completed out of sight. From the north it is an ugly peaked mountain with precipices falling steeply to the sea, and from the south-west it is a pleasant cone-shaped hill of fine lines.

South of An Teallach is Strath na Sheallag, the wide confluence of two glens and their rivers at the head of Loch na Sheallag. The strath is an impressive place, flanked by the shapely peaks of Beinn Dearg, Beinn a' Chlaidhemh, and An Teallach. It is integral in the lives of the deer that roam again over a glen that has a long human history now almost lost. Strath na Sheallag – Strath na Sealga, 'the glen of the hunter'; Shenavall – Seana bhaile, 'the old town'; Larachantivore – Larach an Tigh Mhor, 'the foundations of the big house': the names stand for a past age, with a few old men's stories.

I am not the only man whose imagination has been raised to a state of sensibility by a first glimpse of Strath na Sheallag, coming over the unmarked track from Carn na Carnach. Men and animals love this remote strath, the deer linger on the bog between the rivers till late in June, the snipe drums high above the bog in the early mornings then, Highland cattle come up from Gruinard, and Highland ponies will gather there in May from a range of 12 or 15 miles if the chance occurs. What are the reasons for the charm some places have for beasts and men? Shelter and a fresh bite are not a sufficient explanation. I return to the question again later, but with no full and satisfactory answer. In times of heavy snow there may be as many as six or seven hundred deer in the strath, for they come in there from many places outside my beat.

The south-western face of An Teallach, which drops over 3,000 feet to Loch na Sheallag, is more rock than grazing and is hard-weathered. It has only one corrie, Coir' a' Ghamhna, which may hold 50 stags one summer's day and none the next. Obviously a steep face of this kind has not the holding power for deer that have places like Coire Mor and the Toll Lochan corrie. This ground is grazed by stags only, and they leave it for the strath in bad weather. The lie of the land affects – to borrow Durkheim's phrase – the social morphology of the herds, a subject which will call for fuller treatment in the chapter on territory.

The south-western side of Strath na Sheallag is flanked by Beinn Dearg Mhor, a mountain which stands alone for beauty of outline in this countryside. It is of the same formation as An Teallach, laid and weathered into similar shape, but Beinn Dearg is of lesser bulk and height, standing in a fine isolation and wholly within the field of the eye from across the strath. Its symmetry is almost perfect thus seen from the slopes of Sail Liath. The grazing on Beinn Dearg is much better than on An Teallach, and it holds a greater stable population of deer. Heather, grasses, and sedges are stronger and closer, but there is no change of species. Although on Beinn Dearg the rainfall is as high as and possibly higher than on An Teallach, the ground is drier because the slopes have good drainage gullies, and on the north and north-east are not terraced so markedly as those of An Teallach. Beinn Dearg Mhor and Beinn Dearg Bheag are outposts of the Torridonian Sandstone alongside an area of the archaic Lewisian Gneiss. It is when Beinn Dearg, An Teallach, Sgurr Fheoin are seen from the grey country of the gneiss in the afternoon sun that their names take on their significance; then they are red indeed – red as the hearth – red as wine. When the snow is down, an east wind blowing hard, the sky leaden, and the tops partly hidden, Beinn Dearg and An Teallach roar to one another from the unapproachable country of their summits. I do not know what causes this deep song in the high hills during the weather I have outlined. It cannot be explained away purely on the basis of wind and rock surfaces, for the roaring should be heard then under other sets of conditions which included high wind. I am inclined to place this roaring in the same category of sounds as the phenomenon of the singing sands.

The sharp demarcation at about 1,700 feet and between the dry porous upper faces of the Torridonian Sandstone and the exceedingly wet lower slopes has a far-reaching effect on the herbage and on the winter snow-line, and indirectly therefore on the movements of the deer. These lower slopes are terraced along the contours, drainage lines become devious and obscured, and shallow troughs of waterlogged peat are formed on these horizontal terraces. They never become dry even in summer, and in wet weather it is impossible to get about dry shod. Here the white heads of the bog cotton bob cheerily in the wind of a June day. Conditions on the area of the Lewisian Gneiss, which stretches northward from the feet of Beinn Dearg, Sgurr Ruadh, and Sail Mhor to the sea of Gruinard Bay, are quite different; that country seems another world.

The Lewisian Gneiss, the oldest rock in Scotland, is hard and unyielding. It forms the rugged coast of Gruinard Bay, reaching a height there near Gruinard House of 619 feet in Cam na h' Aire. This dome of archaean rock is typical of many as rugged and much higher inland, for the heights of the gneiss hills, rocky and round-topped, increase farther from the sea until an ultimate height is reached in A' Mhaighdean, 2,850 feet, 10 miles away as the crow flies, and off my beat. The country of the gneiss is a maze of small hills, deep little winding glens, and many small lochans – a turmoil of rock and bog on which the glaciers have left only their scars and rounded boulders of Torridonian Sandstone lying like resting deer on the tops. Man has barely scratched his impression.

Herbage on this ground is more grass than heather – bents and sweet vernal with moss and lichen running through it all – yet on the islands in the lochans and here and there on a dry rock face there is a flourish of heather and blaeberry, with a rowan or a stunted poplar struggling for a living. I have considered the presence of these poplars, old beyond the life of their kind and mere bushes growing where the deer cannot reach them. Even here they do not grow unhindered, for I have seen the few leaves stripped by puss-moth caterpillars. The poplars are isolated, one to a square mile perhaps, relics of a golden age of pine forest and softwood scrub. On large areas of the gneiss they alone remain, with a still more occasional pine.

The lochans on the gneiss formation are peat-laden, the smaller ones are often not more than three feet deep between the surface and the peat, but the depth of soft peat beneath must be great. The deer avoid these lochans for the death traps they are to an unwary beast. But where a lochan has even a short stretch of sandy shore, like Lochan Gaineamhaich, the head of Loch Ghiubhsachain, and Loch a' Mhadaidh Mor, the deer come there to play, wading in the shallows, rolling and scampering along the sand in abandon. The dark lochans of the gneiss grow mare's-tail, water-lilies, and other aquatic plants, and are often full of small trout. The black-throated diver nests here, the goosander fishes at remarkable speed, whooper swans spend the winter feeding on the under-water vegetation, and in summer there are two pairs of greenshanks on this ground for every one on the Torridonian. Many and happy are the days and nights I have spent on this ground. How often have I been wakened at dawn in May and June by the piping of the ring ouzel, the harbinger of each new day!

The lower and north-western part of Gruinard Forest is on the gneiss, the two main hills being Carn nam Buailtean and Carn na Beiste, separated by the flat bog of the Allt Creag Odhar. Across the Gruinard River as far as the Little Gruinard River, which is my boundary, the gneiss becomes still more broken and landmarks are few until the wild glen of the Uisge Toll a' Mhadaidh is reached. On the one side are the three rock masses of Creag-mheall Beag, Creag-mheall Meadhonach, and Creag-mheall Mor, with passes between. Beinn a' Chaisgein Beag is on the other side. The northeastern face of this hill overlooking the glen is a mass of fallen rocks of great size, but the head of the hollow in the hill is smooth. The western face, sloping gently to the Fionn Loch, is well covered with grass and heather, and this hill is a favourite haunt of deer in winter and summer. Its hinds are the largest of any on my beat, and the stags cast their antlers early in the season. At the head of the Uisge Toll a' Mhadaidh is the lonely loch of the same name, bounded to the south and west by louring cliffs of gneiss. The loch lies within the sanctuary of Letterewe Forest and is rarely visited. Deer are more numerous round here than elsewhere on this ground.

Climbing from Loch Toll a' Mhadaidh round the isolated Lochan a' Bearta, Caiseamheall is reached, the steep rampart of the gneiss against the Torridonian. Beinn Dearg Mhor comes red to the eye, and the gneiss piles itself like a grey wave against Beinn Dearg Bheag. Strath Beinn Dearg, remote and deer-haunted, lies below, with Loch Beinn Dearg and Loch Ghiubhsachain at head and foot. In June the glen is fragrant with wild thyme. The hills rise so steeply from the shores of Loch Beinn Dearg that two men, one on either side, can talk to each other without great effort on a still night. One evening in early May I was fishing for my breakfast here after the sun had fallen. The country was still. A pair of black-throated divers were resting on the water a hundred yards away, unafraid of me. Then they rose to go back to their nesting loch, and the acoustic quality of the place enabled me to hear the sound of their movements, amplified but clean-cut. The wings beat on the water sharply for 70 yards until they were in the air. Their streamlined shapes circled the loch three times, making height, and their wings whistled loud in short staccato rhythm. The silence followed. When a hind barks here, the deer of two hills are alert.

The herbage floor of the gneiss is similar whether at 500 or 2,000 feet. The water is caught up in small pools amongst the rock and peat of the rounded summits which contrast so sharply with the spires of the Torridonian. Apart from wind and temperature, the conditions for plant growth at these different heights vary little. Deer may be found high in winter and summer, and the snowline is not one of sharp demarcation. On the top of Caiseamheall at over 2,000 feet I have found a pool in the peat containing palmated newts, a water-beetle larva like a water scorpion, and its long-oared parent too. This would never happen on the Torridonian. Bogwood is plentiful on the gneiss, especially in the glens. The peat has fallen away from a long trunk of it near the summit of Carn nam Buailtean, and there are many roots of bogwood as high as on Caiseamheall and round Lochan a' Bearta. The wide forests of which the bog fir is the remnant followed the last glacial period, themselves succumbing to the sphagnum growth of a pluvial period. This moss formed the recent peat which is now showing early signs of decay.

The country and the weather are an indivisible whole. The weather helps to fashion the country and its life, and the country influences the weather. In this chapter I give only a general outline of its course. Details of temperature, humidity, and wind, with their effects on the behaviour of the deer, come later. Changeability is the weather's main characteristic in the West Highlands, together with the general high humidity. This north-west coast is the principal storm track of the British Isles, and there are more thunderstorms in winter, associated with heavy hail showers, than there are in summer. The name of An Teallach, 'the hearth' or 'the forge', is thought by some to refer to the lightning which plays round the mountain during these storms. Wind is mainly from the south-west; there is often much bad north-west weather in April and the back end of the year, and the steady south wind which blows in the early months of the year is cold after its passage over the snow-capped plateau of the Central Highlands. The mean temperature throughout the year follows a smooth and almost symmetrical modal curve from about 38° F in January to 55–56° F in July, and back to 38° F again in December. The maximum shade temperature I have recorded at Brae House has been 93° F and the minimum 18° F. Sunshine shows a peaked mode of five and a half hours a day average in May–June, down to less than an hour in January and December. The sunshine has an intense quality, however, and the sun of June can produce a deeper tan here than in the Lowlands.

Rainfall is an important factor in this country's weather and its vagaries are interesting. There are comparatively few days in the year without rain, and yet how rare are the days of steady, continual, windless rain! It comes in boisterous showers or as soft mist round the tops, and if there is rain for several days without ceasing, wild weather accompanies it and the monotony is relieved. The rainfall varies widely in amount from place to place within a radius of a few miles. At Brae House, on the northern side of the Dundonnell strath, the rainfall has been 72 inches in the year, but there is much more across the glen. Mr Donald MacDonald kept a gauge for me through a year at Larachantivore in Strath na Sheallag, where the fall was 100 inches. About 65 inches are recorded at Gruinard on the sea coast; and Ullapool, a few miles away on Loch Broom, has an annual average of only 38 inches. The distribution of rainfall through the months of the year follows the somewhat irregular curve for the rest of western Britain. October, January, and February normally give the highest figures; and March, June, and part of November are the dry months. The rain and snow, when they come, tend to travel about in patches. I have lain on a hillside in sunshine watching half a dozen showers seeking their way in and out of the mountains. And from Sail Liath I have seen the hail and snow showers advancing slowly and silently up Strath na Sheallag in grey, crescented columns, imperturbable and inexorable. Once I saw an eagle soaring in the van of one of these storms with a fine contempt. Perhaps she rose above it as it broke in a swirl round me and other earth-bound creatures, for she was still resting on the air above Beinn Dearg when the squall had passed on up Gleann an Nid. The play of rain and storm in this country makes for its changing beauty of light and shade. Each day, through every season, is a pageant of colour, whether the crystal hard light of sunshine on the rocks and water after rain, or the vivid shimmering greens and blues of summer. There is a gold in September which is of the sunshine itself as well as in the dying leaf. There are velvet textures in the browns of bracken and sedge and the purple of leafless birches when, on a winter's afternoon, the air is clear and still under a low mist, each twig bejewelled by hanging drops. Few days are all grey and colourless. August is the most unpleasant month of the Highland year, with much rain, hot nights, and little wind; mist is a hindrance throughout the month, and telescopes and binoculars give a dim field. Sometimes in winter after a good spell of weather there will be a day of biting cold and strong south wind under a dark sky, which presages storms to come. The deer are irritable and restless, and the irritability extends to the human observer. How welcome then is a shelter behind a rock until the cold drives to action once more!

Atmospheric relative humidity is usually high, and variable from minute to minute. In dry, warm weather it may fall well below 50 per cent, and in March, when a steady but slight east wind blows and the skies are clear, it is down to 20 per cent. Then the dry sedge sings in the breeze and it is crisp underfoot. Cold, wet weather with the usual wind seems very cold here, but actually the winters are mild. Snow does not come down to the sea's edge more than once or twice in the year and then usually to disappear in a few hours, or at longest a day or two. Fourteen degrees of frost have been the most I have recorded at Brae House. In December 1935 there was a continuous period of three weeks of frost on a light snow. It went hard with the deer. Heavy snow may come as the culmination of wild weather in early November or early April, and fine settled weather with good frosts follows for a few days. Travel on the hill is impossible until the snow freezes hard, but unlike the Central Highland districts, the snow does not often freeze hard enough to carry a man until the 1,750 foot level is reached. The deer are in similar straits, and their journeys of several miles over frozen snow which may be traced in forests farther inland are consequently uncommon in the western forests. When, however, the surface is good for a day or two, men and animals find joy in movement and the cold is not felt by them in the dry, still atmosphere. Thaws are apt to be sudden, and then irritability becomes apparent in the herds. A sudden and complete thaw is welcome to man and beast, but a cold thaw causes much discomfort.

Daily maximum and minimum temperatures vary most widely in May and June, with what effects on the movements of the deer will be recorded later. In August and early September there is often morning mist in the glens away from the sea, and the hill-tops stand out in the clear air above for an hour or two as blue islands in a white and silent sea. At such times the early morning is the best hour of the day, and the hill-top the best place. Then the temperature on the hills is higher during the night than it is below. There is a period of little change between day and night temperatures during November when the deer go high.

All these conditions are reflected in the movements of deer to a greater or lesser extent, until the close observer may be able either to predict the coming weather from them or, conversely, to know where the deer will be from the course of the weather. The surer course is to forecast the weather from the doings of the animals.

Here and there in this chapter I have allowed myself to depart from purely objective description of the country; usually where my impressions have been particularly vivid, as for example in Strath na Sheallag. I yield again now at the conclusion of this chapter to the same temptation. Individual hills and glens have their own characters, and rivers achieve almost a personality in the imagination. I think the area of the gneiss in its aloofness and lack of outstanding landmarks has made the deepest impression. Nowhere have I felt more the ephemeral nature of individual man than after spending some days alone in this grey, broken country. I have lain sometimes on the western slopes of Beinn a' Chaisgein Beag, 'the hill of cheese', which are rich and pleasant and where, doubtless, man's animals have grazed in past times. The burns fall to the waters of the Fionn Loch, gleaming as white as its name in the June sun, and there are traces of the dwellings of men. I have heard the singing of women's voices and the laughter of little children in this place. Perhaps the play of wind and falling water made these sounds – I neither know nor care – I was content to listen in the beauty of the moment.

These strange qualities of this part of the country, inviting or repelling, are real to men. Under Beinn a' Chaisgein Beag on the shores of the Fionn Loch they are happy, but at the head of Uisge

Toll a' Mhadaidh the scene has changed. The cliffs fall steep to the loch and the ground about is as rough as could be with fallen rocks, deep peat hags, and heaped moraines. I have found it a strange place, and the same thought has been murmured to me by some of the few men of the country who have been there. These sensations may be caused through the eye by the dispositions of masses and planes and their relation to the course of the sun, as well as by the huge rock surfaces devoid of vegetation. There are many such places in the Scottish Highlands where seasoned men – myself too – have had to move out at nightfall. The sensation is not fear, for intimate knowledge of the place disposes of that; but there is discomfort sufficient to make a man move. These problems of the character of individual places must remain.

The country as a whole has a joyful quality, and the constant change of lights and shades to which I have referred are stimulating to the seeing eye. I have not fought the country these two years but have let it be my foster-mother. Her discipline has been stern but her smile is never far away.

CHAPTER TWO

Technique and Personal Reactions

Equipment

WORK OF THE NATURE described in this book calls for good and carefully chosen equipment. The person has to be clothed in such a manner that he will not clash with the landscape. Lovat and *crotal* mixtures of Harris tweed fill this need. Khaki handkerchiefs were chosen in preference to white ones, and plenty of them carried, for the nose runs more than normally when moving about on the hill. As long periods have to be spent lying on the ground, often during inclement weather, some form of windproof clothing is very helpful. I had made a suit of Grenfell cloth, double thickness, fashioned on the pattern of the porters' suits in Everest expeditions. The trousers and hooded tunic have no open seams, and as the tunic can be pushed inside the trousers, and the whole suit weighs only four lb, I was lightly clad and without any ragged ends such as ordinary suits make. The colour is olive and fits in well with the herbage floor. It is absolutely windproof. An observer gets wet through inevitably in this kind of work and that has to be endured. Waterproofs should be avoided completely. The Grenfell suit has one serious disadvantage for me. The material, which is a finely woven cotton twill, makes a swishing sound when drawn against itself or against the herbage, and the deer hear it. Harris tweed is almost silent. Heavy shoes and thick woollen stockings (hand-knitted) are necessary for long journeys, and I had the insteps of my shoes fitted with Tricouni nails. These allow one to cross the burns more easily, for a leap from rock to rock can be made with more assurance. The little hard teeth of steel bite on to the rock most efficiently.

Messrs Thomas Black and Sons, of Greenock, made me a light tent, nine feet nine inches by five feet six inches, in olive-green material. It weighed less than 14 lb together with a large groundsheet and it stood up to very bad weather. From a short distance away it was almost invisible, and the deer have come to graze within 200 yards of it unheeding.

The best of telescopes only are good enough. Mine is by Ross, two-inch objective; x 20. This glass gives a large, well-lighted field and the images are dead sharp. A higher-powered glass than this is a doubtful advantage in the Highlands, for the light is often bad and a high power gives a misty field and an indistinct image. My glass is in aluminium, very light and handy, and carefully used by one man this material is entirely satisfactory for the hard, all-the-year-round work I have given it. I carried also a pair of Voigtlander 32 mm x eight binoculars for short distance and quick observation. This particular pair of glasses was the envy of the countryside and I would not be without them, although at one time I was rather contemptuous of binoculars. These glasses, which give a fine stereoscopic field, are useful for scanning ground roughly at a distance or very carefully near at hand. An observer of deer or similar animals should not fall into the error of thinking that ground can be thoroughly scanned through binoculars. The field looks good but the eye misses many animals. The telescope only should be used for this purpose, the ground being covered inch by inch.

My camera is a Leica, and with its range of lenses, automatic focussing, and small size it seems to me there is no better instrument for this type of work. You cannot carry a large camera many miles on the hill, and certainly it is impossible to stalk with anything large and box-like. All the photographs in this book were taken with the Leica. Photography of deer on Highland hills has been a great disappointment. The animals blend so perfectly into the background that the monochrome print shows no contrast. Even with a 13.5 cm long-distance lens a photograph is not satisfactory taken at beyond 50 yards' range. Too much stalking to that distance frightens the deer and upsets the main course of work and behaviour. My aim was always to stalk *away* from the deer as well as up to them, so that they should not be disturbed. The backward trip is much the harder. The quality of light for photography of the deer is often so poor that I have had almost clear films (Agfa Super-

Panchromatic) using 1/20th second exposure at a stop of f. 4.5 – and this near midday.

Stalking

The first necessity is to become familiar with the ground, a task which takes time. Management of the wind when approaching deer is of very great importance and links up with intimate knowledge of the country. The 'carry' of the wind as seen by the passage of clouds over the sky is not always the direction it will be taking in the corries. Indeed, it may be the opposite. These conditions have to be remembered, but there are means of estimating wind in places ahead by watching through the glass the herbage bending, the way in which the hair of the deer's coats is blown, and the flight of birds. In sunny weather there is not only oneself to keep out of sight but the shadow too, and this can be an awkward appendage. Light is of great importance in stalking. If the sun is not shining and the light is diffuse, the hazards may be considered equal, but bright sunshine – especially after rain when there is some cloud about – can weight the chances. This sun behind you makes stalking easy; but if in front, approach may be impossible. The advantages are with the deer, and the sharp lights and shades usually prevent the observer from seeing *all* the deer which are on the ground ahead. It is most important that each new bit of ground coming within the field of the eye should be scanned with the glass before moving on. Single animals have a disconcerting habit of grazing apart from the main body of the herd, and if these are startled the stalk is finished.

It is worth while getting to know a fair number of points of vantage from which a large area of ground can be scanned. Long periods are spent in these places watching the day-to-day movements of the deer, and an idea of territory is gained. I make a habit of counting deer whenever I see them. Estimation of numbers of animals is very difficult – there are usually less than you imagine – and although I may be watching one lot of deer for several hours

I count them every few minutes, so that those behind knolls and otherwise missed are ultimately included in the tally. The whole of the forest should not be traversed and combed with a view to making a census until the groupings and favoured grounds of the deer are known. A notebook should be carried, and however irksome it is to take notes *at the time*, this should be done if possible. Furthermore, all actions and patterns of behaviour should be timed. On certain days of the year – such as, approximately, 15 July, when the deer have gone up to stay; 15 October, in the height of the rut; 15 November, when only a few young stags are rutting and the harems broken up; 15 March, if the weather is good and the deer are wandering; 15 June, in the calving season – I have made special journeys through as much of the ground as possible in one day. I attempted no stalking on these occasions, for the aim was to see where as many deer as possible were at one time. By using the high ridges as much as I could time and distance were saved, and speed was essential. Such days mean 35–40 miles of walking and 7,000–10,000 feet of climbing, and they are among my most pleasant memories.

During some years of laboratory work I had not noticed the deterioration in my ability to see animals against a hillside. I realized it only by the improvement which took place in a few weeks when I began this work. If you are to develop this quality of sight so that deer and other objects stand out from the hill in your field of vision, not much reading should be undertaken. This seems to alter focal habit, and a hillside, instead of having depth, becomes merely a flat homogeneous expanse.

How can one class of deer be distinguished from another? This power to distinguish is necessary if the social system and the dynamics of population are to be understood and interpreted. It takes time for the eye to become accustomed to recognize differences, and once that has occurred the nature of the differences has to be defined in the mind by careful self-interrogation if the matter is to be set down on paper. There are some things which escape the scientific approach: for example, a deer may be seen 600 yards or half a mile away, not on the skyline but against the hill with which its colour blends so well. Is it a stag or a hind? Probably the antlers will not be visible at that distance unless the beast turns its head and there is a glint from their outer edge – and in April the antlers will be shed. Yet the experienced observer will know well enough to be right nine times out of 10. The size of the beast and possibly the shape of the neck are the likely clues. The difference is one of a few inches and the distance of the animal from the eye considerable. This seems obvious on paper, but it is quite a different problem on the hill. The stag's ears are not carried so high as those of the hind, and when he runs his neck seems to be nearer the perpendicular. This may be just illusion caused by the different shapes of the neck, but the experienced watcher knows he is right before he focusses his glass. An inexperienced observer will pull out his glass to look at a rock as big as a house, thinking it is a deer. Another problem at 600 yards is whether the animal has its head up looking at the observer, or is still grazing. I cannot give definite reasons for knowing, for no sharp lines can be seen. My own feeling is that, the position of the legs and the curve of the back being different in each position, the eye is able to discern the patterns, however indistinctly; but the critical mind behind balks at the flimsy evidence. Is the eye too quick for the mind? Perhaps this is the stuff of intuition. The fact remains that an observer has to go through a period of conditioning of a most subtle kind; and, question himself as he may, he will be able to give no better answer than the plain, correct, but unsatisfactory one 'I just know it is so.'

Let us suppose now that we have a group of hinds below us against a good green background for observation. What are the differences between a young maiden hind, a yeld[1] hind, and a milk hind? The observer has to be in an unusually favourable position to be able to see the condition of the udder. Of course, the calf of the year may be grazing near enough to identify it with its mother, and it may be seen actually sucking milk. But it is often necessary to decide on the constitution of groups without these aids. At first the observer is apt to give the same answer as in the last paragraph,

---

[1] Barren, or not with a calf at foot.

but the subjects are really much more amenable to definition. There is no difficulty about yearlings because they are still smaller than adults and their heads appear shorter compared with those of the hinds. The two- and three-year-old hinds still have this short quality about the head, but to a much less marked degree. The head of the adult hind is long, well dished like that of a Jersey cow, and very lean. Her neck is also very long and she is obviously ewe-necked. I say obviously because the hair is short and the character fully exposed. Yeld and young hinds have slightly longer hair about the neck and the illusion is that the neck is shorter and not so concave on its upper surface. Colour is another rather fickle aid for distinguishing milk hinds from yeld hinds. Some hinds have a good deal of black between the eyes and down the fore face, and there is a black streak running down the neck and along the back. There also may be something of a transverse cross of black over the chine, and the front of the knees and metacarpals are shiny black. A fine ruddy colour of the rest of the coat usually goes with this black marking. Such a hind is almost certain not to have a sucking calf. Lactation seems to bring about paleness in the coat and the black may go rusty or fade. It must be remembered that red deer vary widely in colour from buff through dun to the dark characteristic red, and some hinds never have the beautiful black marking I have described.

It is very difficult for an active mind stuffed with the matter of 'education' to play its part effectively in stalking animals. Such an observer has to spend long periods walking or, in one sense of the phrase, doing nothing. An idea strikes him at these times and his mind begins to work on it. That is fatal to the task in hand, for suddenly he sees deer moving off or one animal galloping to frighten a herd farther on. The observer must empty his mind and be receptive only of the deer and the signs of the country. This is quite severe discipline, calling for time and practice.

If you are going to observe an animal well you must know it well, and this statement is not such a glimpse of the obvious as it appears at first. It is necessary intellectually to soak in the environmental complex of the animal to be studied until you have a facility with it which keeps you as it were one move ahead. You must become *intimate* with the animal. As I read Jennings's *Behaviour of the Lower Organisms* I feel that he has achieved that state with his *Paramoecium*, a much more difficult task than I have had in living near to an animal which exhibits emotions which, I must conclude, are not far removed from my own. I would emphasize the importance of thinking of little else but the animal and its environment until one's intellectual complex has become 'tuned in' on them. In this state the observer learns more than he realizes. Sometimes I would go for several days in succession without seeing anything of particular interest. Very pleasant, I might say to myself each evening, but what new point has come from to-day's work? None, I would admit; but at the end of a week of such days something had been learnt. The pattern would be clearer somewhere, or a new idea would have called for special observation in the future.

I have been interested to note the reactions of my own senses. They all sharpened, and I realized as never before how they all work together as a complex. On one occasion I caught a cold which temporarily took away my sense of smell and taste. I found my ability in stalking to be much impaired. Sight and hearing, the two senses which I had thought to be chiefly used, were not enough. The Grenfell suit has a hood and a zip fastener which keeps it close round the face. Often enough the weather called for its use, but I found I could not stalk well if my ears were covered, so the hood was hardly ever used. The capacity for awareness seemed to be lowered although I had not consciously been listening to or for anything. During the summer of 1935 I went barefoot, and after a fortnight of discomfort I had my reward. The whole threshold of awareness was raised, I was never fatigued, and stalking became very much easier. This ease in approaching animals was something more than what was gained by leaving off heavy and possibly noisy shoes. The whole organism worked in better co-ordination.

In concluding this chapter on ways and means I must admit that these two years have brought the deer very near to me, although I have had a lifelong love of them. These creatures are more than the material of the scientist's paper. They are animals whose lot has been closely linked with human development. We have pitted our wits against them through thousands of years, and the hunter is not worth his salt who does not admire this quarry and is not content sometimes to watch the beauty of their lives, free from the desire to kill. I have had the best of it – the love of them, neither wish nor necessity to do them harm, and a long time to watch through all seasons of the year.

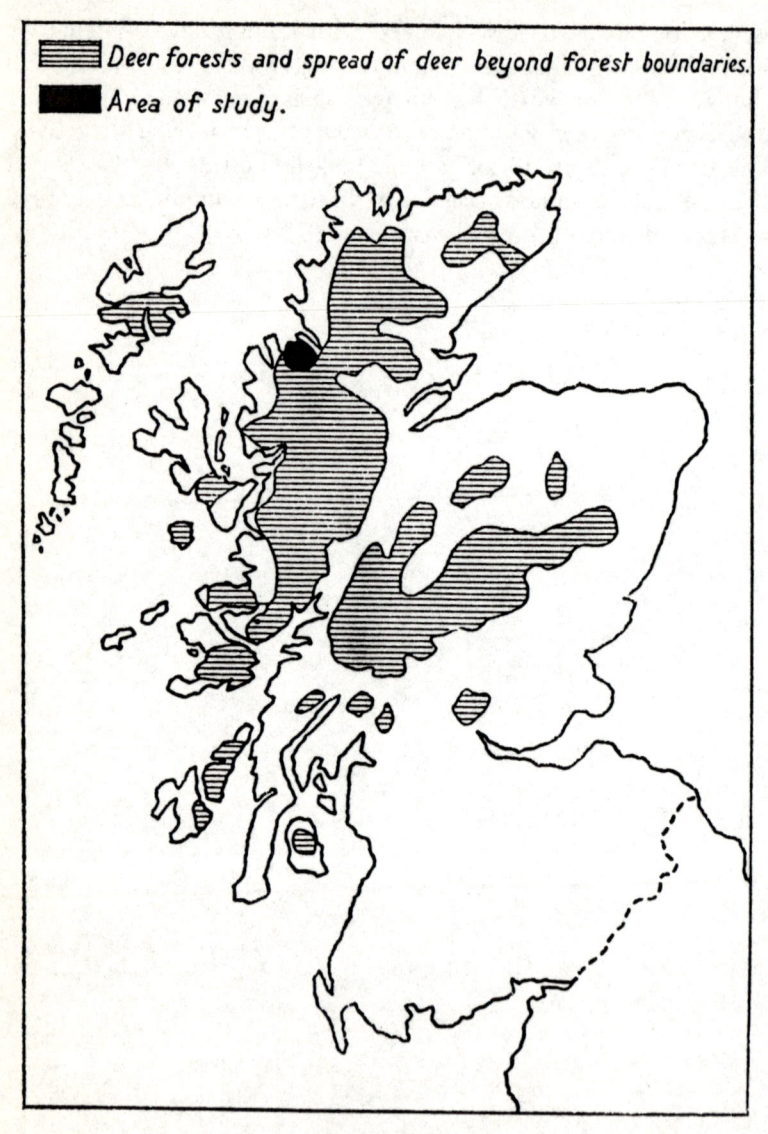

MAP 2
Distribution of red deer in Scotland

CHAPTER THREE

Territory and population

FREE-MOVING ANIMALS need space in which to feed and breed, to rest and play. Conservatism of habit, a factor of importance for the survival of species, tends to restrict movement to a particular area. True nomads, defined as creatures which wander fortuitously and have no home ground, are rare in nature. Choice is another reason for individuals or groups remaining on one area. Animals live in definite places because they like them. Familiarity with one piece of ground enables an animal to use it in the most advantageous manner for its comfort and well-being.

Territory should be understood as a different concept from range. The one is concerned with the intimate life of individuals and groups; and slight incidents, psychological idiosyncrasies, and facts of being loom large in territorial affairs. Range is a much more impersonal matter dealing with the geographical distribution of a species, the alimental and other main environmental limitations which modify the area of diffusion. Science has gathered together a large body of accurate knowledge concerning range or geographical distribution and its fluctuations, which has had an inferential value in philosophical biology as great as that of its intrinsic facts. Indeed, the study of range has tended to obscure the importance of observing territory within range. Data are meagre on this subject as yet, and though the animal ecologist is trying to make good the deficiency, it seems that the main impetus to the study of territorial observance will come from an increasing desire to understand animal behaviour. Howard's classic work (1907–14, 1920, 1929), for example, on the breeding territories of birds shows a strong psychological bias. Nesting, however, is only one phase of the bird's year, and the bird by its very mobility is difficult to observe territorially except at nesting time. The subject has been studied in greater detail in insects, and I need go no further than mention the work of Wheeler (1928); and Elton's paper (1932) on the discrete territories of ant communities. In mammals it remains a field of opportunity for the inquisitive mind.

Territory is more difficult to observe and define than range, for it requires close attention to particular groups of animals throughout the seasons of one or more years. The human watcher is hampered further in this task by his physical body, to the presence of which most of the higher animals object. They are elusive and usually they can get over the ground better than man. To keep out of sight or immediate scent is not enough, for our tracks retain our scent for many hours, and unless the observer uses forethought and is skilful in his movements he may not gain any knowledge of territory, or only a distorted idea. Work on territory in the higher animals leads us into a wide field of biological thought, particularly that concerned with the development of sociality. Ecology, the study of the organism in relation to its environment, is on the one side; the study of territory, and social psychology with its ramifications, are on the other. Socio-ecology as a subject has, as I see it, a definite function in biology. It can serve, amongst other things, as an excellent check on over-specialization and the drift to the laboratory.

Range, then, is determined physiologically more than psychologically; territory psychologically as much as, or more than, physiologically. There are two great classes of territorial animals – hunters and grazers – and this division is of greater significance to the observer of animal behaviour than in zoological classification. The hunting pack is comparable with a capitalist organization having a fair measure of state control. Territory is guarded jealously and fighting follows intrusion. Size of territory is governed largely by the density of numbers of the animals hunted. The limited degree of gregariousness and co-operation is for the purpose of hunting rather than being familial in origin. In fact, many hunting animals show no sociality beyond the single family. The hunting animal which invades another's territory goes in fear of his own kind and is ready to run or fight. The grazing herd, on the other hand, shows a state comparable with the City of the *Laws* of Plato: communality and disciplined orderliness which need but little discipline. Environmental conditions being equal, territories are bounded by choice and not by jealousy. If circumstances call for it, another group's territory may be occupied; but this occupation is tolerated, and the groups may combine in face of the common need. The herd is in itself a bulwark against the hunter and predator.

The red deer, *Cervus elaphus* L., in their social life appear to me to have reached the highest development of sociality of the grazing herd. The persistence of the species is dependent upon it. Their social system is a matriarchy, founded on the family. The sexes separate into their own groups for the greater part of the year, observing the boundaries of their own territories. Without going deeper at the moment into details of social structure, from this point we can continue the study of territory in red deer and of the particular herds which I have watched.

The territories of red deer may be divided into three classes:

1. *Winter Territory*. This ground is sharply defined on three sides, the fourth leading to the summer grazings and therefore variable. Winter grazings are for the most part on the lower slopes of the hills between sea-level and 1,700 feet. The winter territories of hinds and their followers constitute what is embodied in that rather nebulous phrase used by Seton of North American mammals – 'home ground'. Home, as I understand it here, means that the winter territory is also the calving ground of the hinds in June and the place where they gather in October when the rutting season occurs. The winter territory of stags appears more as a refuge in stormy weather and cannot be given the same value of home as in the hind groups.

2. *Summer Territory*. This is contiguous to the winter grounds and extends to the hill-tops. It will be obvious that the area of the lower slopes of a cone-shaped hill to a height, say, of 1,500 feet must be greater than the area above that contour which includes the summit at 3,000–3,500 feet. If summer territories of each group occupying winter grounds at the foot were to complete accurately the sectors to the summit of the hill, the shapes would not be those of probable animal territories. Summer grounds which include the summits are, in fact, grazed communally to a large extent, but the groups retain

their own social integrity. If climatic conditions call for it, each group will fall back on to its own winter territory.

3 *Breeding or Rutting Territories.* There is a certain artificial character about these territories. They are the areas upon which a stag can herd together conveniently a number of hinds as a harem, and their primary relation is not to the herbage quality and those other factors which influence deer in their choice of winter and summer territories. The rutting grounds of the stags are of space rather than of place; they are always on the territories of the hinds, that is, the stags come in from their own summer grazings towards the end of September, and they are grouped together on well-defined portions of the hinds' territories. One stag's rutting ground may be 25 acres or five, depending upon the earliness or lateness of the season, whether he has 50 hinds or five. A stag cannot keep together 50 hinds except at the very beginning of the breeding season, so that at the height of the rutting season of six weeks the areas of individual stags' territories are at their smallest. Each stag and his harem on an area of 250 acres may occupy any part of it for a while, though for the time being his particular territory will be small. Boundaries must be arbitrary and subject to considerable alteration, but they are exceptionally sharply defined psychologically at any single moment. These territories are held by the intensity of the stag's sexual jealousy. Why should the rutting territories be grouped on the hinds' grounds in close proximity to each other? If MAP 3 is considered in relation to MAP 6 it will be seen that they are situated on some of the less steep ground. That is as far as the map can help, but familiarity with the place would show further that the ground of the rutting territories is comparatively easy for a stag to scan and to run over. Obviously this is a desirable condition for an animal which keeps its territory by its own wits and activity. The hinds are unconcerned and tend to wander to the rougher ground. The stag herds them back and there is an evident tendency for him to work the hinds downhill rather than uphill, a task in which he never quite succeeds. Given the fact then, that rutting territories

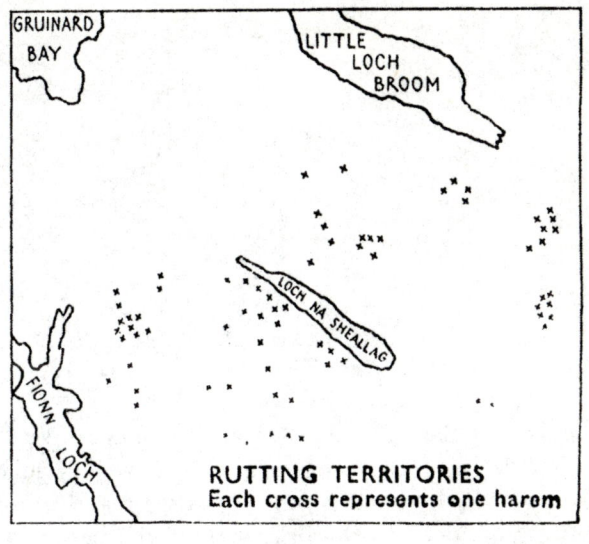

[Map 3: Rutting Territories. Each cross represents one harem. Shows Gruinard Bay, Little Loch Broom, Fionn Loch, and Loch na Sheallag.]

MAP 3

are on the hinds' grounds, it is the stag who definitely works to keep his hinds on ground which allows him fair mobility. Another reason which keeps the small rutting territories in a state of flux is the challenge of fresh stags and the departure of spent ones. At the beginning of November the rutting territories begin to disintegrate; the rut may not be at an end, but the older stags are spent and the younger ones are amongst the hinds, and, as I shall show later, young stags have not the finesse or desire to establish well-defined rutting territories.

Before studying in detail the territories on my ground something should be known of the distribution of the population. I have shown on MAPS 4 and 5 by means of crosses the numerical distribution and density over the hind and stag territories. The seasonal movement to and from the upper reaches of the hills should be kept in mind because I have represented the deer as being on their winter territories for the most part – their 'home ground'. The Torridonian Sandstone does not carry as heavy a stock of deer as the Lewisian Gneiss, but the small area of the Eastern Schist – Carn na Carnach

MAP 4

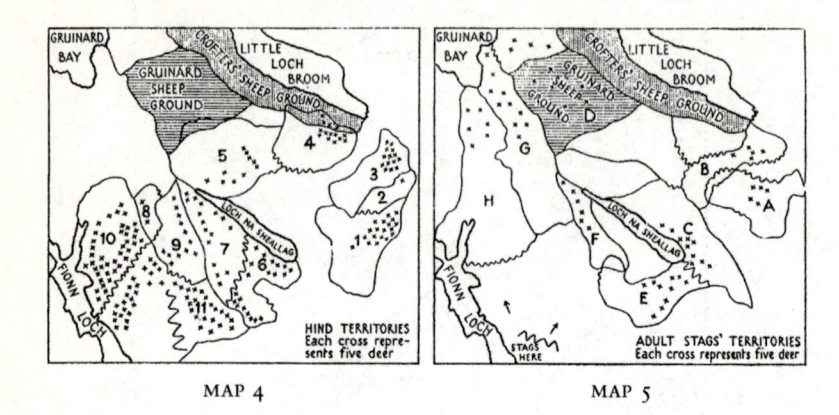

MAP 5

– is heavily populated. Beinn Dearg has more deer than An Teallach, the higher portions of the gneiss are well populated, and the highest density is round Beinn a' Chaisgein and Caiseamheall where, as I mentioned in the first chapter, the best and quietest ground lies. Part of these areas is the Letterewe sanctuary and very rarely disturbed. There are low densities on the lower part of the Gruinard ground round Carn na Beiste and Carn nam Buailtean and along the steep southern shore of Little Loch Broom. The presence of sheep is the reason for the few deer on these last-named areas. Deer and sheep have similar tastes in grazing, and while the carrying-capacity of a forest is lowered by even a light sheep stock, say one to 10 acres, there is disturbance by men and dogs which is, I think, of greater importance. This is kept to a minimum on the Gruinard ground where only one shepherd works the ground with care, but on the crofters' strip along Little Loch Broom, where each crofter may separately visit his few sheep on this common grazing, the deer have become more than ordinarily circumspect, and they are careful not to come so low as to allow a dog to get above them. The strip is not their home ground but the edge of their territory.

A moment's consideration will show that the greater the number of deer in a group of any given density, the larger the area over which each member of the group can wander. This is understandable when we remember that the family is the basis of sociality in red deer. It should be correlated also with what is said in the next chapter and not considered baldly in relation to the areas shown on the plans. Good feeding conditions throughout the year mean more hinds calving annually and less spring mortality among the young stock and old beasts. It may be objected that if the high density is on the better ground, does not a smaller area suffice for food wanderings? We are brought up against a problem here. Good food conditions do influence density, but only to a limited extent. Some species of mammals, like some human types, can live in seething masses whilst food is available; but deer, for all their highly developed sociality, must have plenty of room: they are not among the slum-dwellers of the animal world. In the artificial conditions of deer parks it is possible to raise the density by the constant use of lime, good drainage, and other anti-helminthic measures. Beast for beast, however, the density still remains well below that which would be safe for cattle. Domesticated animals of the farm also react to density of their own species quite apart from the bare total of food available. Good land does not necessarily carry a cow to one acre as compared with one to three acres on poorer ground, but one acre may carry an even higher equivalent of stock of mixed species, all feeding on the same herbage but possibly different parts of the plants. The art of husbandry concerns itself diligently with these assortments, tacitly admitting the principle of density in aspects other than the purely alimental.

Worms play their part in regulating the density of their hosts' populations, but there still remains the psychological factor and this is imperfectly understood. In deer parks where the high artificial density of a deer to five acres may be attained (but under good management is not) accidents are common. Stags fight in season and out, and stags will kill calves and strange hinds under such conditions. In short, overcrowding results in anti-social behaviour which in itself is one type of check to the further increase of a cramped population. Paddocked deer, deprived of the exercise of sociality and a full set of natural conditions, are in still worse plight, and hardly ever look healthy to an eye accustomed to wild deer in the Highlands. I mention these artificial conditions only to stress the fact that as a species red deer react strongly to any imposition of overcrowding. The several kinds of deer as a group, under a wide variety of conditions, exhibit this same desire for space, and it is remarkable how similar are the densities for wild or semi-wild members of this group, whether reindeer in Alaska, white-tail deer in the United States, or red deer in Scotland. The following figures are quoted from Leopold (1933), Hadwen and Palmer (1922), Palmer (1926), and Townsend and Smith (1933):

TABLE I

Species	Locality	Density	Date
White-tail or Virginia deer (Leopold)	Itasca Park, Minnesota, 390 square miles	1 deer to 32 acres	1920
	Grand Island, Michigan, 22 square miles	1 deer to 30 acres	1923
	Pennsylvania forest area, admitted to be over-grazed (Clepper, *cit*. Leopold, says 1 to 25 acres could be sustained in the Pennsylvania area, and that 1 to 40 acres would be conservative stocking)	1 deer to 12.5 acres	1931
Roe deer (Leopold)	France	1 deer to 25 acres	1931
Red deer (Leopold)	Bohemia (heavily wooded conditions)	1 deer to 100 acres	1931
Reindeer (Hadwen and Palmer)	Alaska, semi-wild on ranch system	1 deer to 30 acres	1922
	Alaska, Palmer's later figure	1 deer to 40–50 acres	1926
	Norway	1 deer to 30 acres	1926
White-tail deer (Townsend and Smith)	Adirondacks (about 400 square miles) a decreased stock and heavy woodland	1 deer to 300 acres	1933

Leopold (1933) mentions the slaughter of deer in California rendered necessary by an outbreak of foot-and-mouth disease in 1921. In two years 22,362 deer were cleared off 1,142 square miles. This represents, doubtless, the largest and most accurate census of a deer population ever made and, we may hope, ever likely to be made, for the necessity was of a catastrophic kind. After slight adjustments for incomers, the density over this great tract is given as one deer to 32 acres.

On my ground the density is one deer to about 40 acres, inclusive of water. The density is from one to 30 acres on the Beinn a' Chaisgein Beag and Caiseamheall areas to one to 60 or 100 acres on the northern face of An Teallach and on the Gruinard sheep ground. I am inclined to look on the higher densities as being too high, because there is movement of hinds from Caiseamheall and Beinn a' Chaisgein Beag over Beinn a' Chaisgein Mor in summer and good weather. This hill is off my beat and I did not go over it more than three or four times in two years, for it is part of the Letterewe sanctuary. The toothed lines on MAPS 4 and 5, which show the territories, are intended to indicate that the boundaries in those places are not sharp. If Beinn a' Chaisgein Mor had been within my beat I should have had to include some more hinds, probably about 100, and another 60 or 70 stags which live on the ground at the back of Carnmore.

The population has been assessed by myself only from my own observations in a broken, mountainous country, and I do not wish to state these figures as being absolute; but they are near the truth. It may be pointed out, possibly unnecessarily, that the actual area of ground represented by the map is much greater than that which would be obtained from a simple computation from the map's flat surface. The area in plan of my ground is about 67 square miles, but the real area is over 80 square miles. The steeper and more broken the ground, the greater is the discrepancy. Map-distances are equally deceptive. The correction factor of x 1.5 is a useful one to keep in mind when measuring cross-country distances in the northern Highlands. It is even safer and much more convenient, once the country is familiar, to reckon distances by the number of hours away. The length of Loch na Sheallag may be covered in an hour and a half, but in the Caiseamheall-Creagmheall Mor country one map-mile may take the best part of an hour to traverse.

In the best of faith, estimates of populations of animals from people not scientifically interested are of little value. I have asked many stalkers what they considered to be the density of deer on the forests over which they worked. They never have been able to tell me. Through a long period of time the stalker has learnt to gauge accurately the number of stags which may be shot on the forest each year without depleting the stocks. It is possible from this figure and knowing the area of the forest to gain a rough idea of density. Leopold (1933) shows that for American mule and white-tail deer a 'unit herd' of 24 animals is needed for each stag shot. This 'unit herd' is made up of five stags, five yearlings, seven dry hinds, and seven milk hinds; calves are excluded. The figure is applicable also to Scottish conditions, though the composition of the unit herd may be a little different. About 55 stags are taken from my area each year, which number multiplied by 24 gives a total population of 1,320 theoretically. I have stated already that the density as observed by me is one deer to about 40 acres and that the area is approximately 52,000 acres, which gives a population of 1,300 deer. (My census gives 1,315, but I should not like to be categorical about 15 one way or the other, because I have numbered to the nearest 10 always. Attempts at closer estimates would be more pedantic than accurate. An assessment of numbers of red deer on a rough mainland forest should allow a possible error of 2.5 per cent on each side of the total.) Calculating on these lines, I find that the density of red deer in the best forests rarely exceeds one to 30 acres. Some favourably placed hind forests approach a density of one to 25 acres, but this is perilously near overstocking. It is necessary, of course, when forming this deductive type of estimate to make sure whether the number of stags shot tallies with the number which it is thought could be reasonably taken off the ground. Some forests are consistently and purposely undershot.

The number of 55 stags taken off my beat annually is an average. There may be 50 in one year and 60 in the next. This figure itself provides the possibility of another check when linked up with the age at which a stag becomes prime, say 10 years. Some are shot at this age, some beyond it, but most of them before then, and eight years would be a good average. As the stag companies are composed mostly of stags upwards of three years old, the number of stags over this age on the ground should be at least five times as great as the number taken off each year. In practice the total would be greater because of incidental deaths and wanderings. On my ground then:

5 × 55 + 40 (as margin for all contingencies) = 315.

The type-classes table below shows an annual incoming group of 60 to the total of adult stags, which again allows a margin for deaths other than by shooting during the stalking season.

The number of hinds shot is small, especially in the Beinn a' Chaisgein Beag and Caiseamheall areas where they would be difficult to remove. The effect, as far as depressing the population is concerned, is slight. It is worthy of notice, however, that these areas, undisturbed and unshot, do carry by far the heaviest density of deer.

There is an obvious differential sexual factor affecting the population in that many more males than females are shot. My own figures show 35.6 per cent males to 64.4 per cent females. This distortion of the sex ratio is apparent wherever antlered game is pursued. I believe, too, that there is a slightly greater mortality rate in the male sex, referable to fighting injuries (a very slight influence), exhaustion after the rut if the weather becomes very bad then, and spring starvation after the antlers are cast. Antlers are physiologically expensive, particularly on very poor ground.

Cameron (1923) does not give actual figures of the sex ratio of red deer on Jura, but he says that the proportion of hinds to stags over the island beats varies from three to one to a state of equality. These figures, however, do not justify our taking an average figure of two to one, which would approximate to my own. It is unfortunate that there are no comparable figures which refer to territories within a forest, for a basic understanding of the constitution of group territories is the necessary key for gaining a vivid idea of the dynamics of population and the social life of the animals.

Leopold (1933) records that a tally of 12,531 mule deer in the National Forests of Montana gave 39.0 per cent males and 61.0 per cent females. A 'one buck law' was in effect. He says also that of the 22,362 mule deer killed on the Stanislaus Forest in California owing to the outbreak of foot-and-mouth disease, 48.0 per cent, were males and 52.0 per cent, females. Townsend and Smith (1933), working on white-tail deer in woodland country, give a ratio of 2.5 does to one buck, which is equal to 70 per cent females and 30 per cent males. They admit, however, that the bucks are more shy than the does and do not spend so much time at the feeding grounds, which, of course, renders the buck population more difficult to count.

So long as law and custom do not allow the numerical discrepancy between the sexes to become much wider, there is no fear of productivity in the herds being adversely affected, but the total percentage of males must not be allowed to mask the fact of the much lower percentage of adult males which is responsible for the bulk of the calf crop. From the point of view of leaving a good number of breeding stags on the ground, the number of injured deer should be added to the total kill. Happily this number is small in Scottish forests, certainly not more than one or two per cent, for to leave a wounded stag is breaking an unwritten law. A lame stag does not come into rut in the year of his accident. Leopold (1933) estimates that in parts of the United States the number of cripples is between 20 and 30 per cent of the total kill.

A digression may be justified here in order to consider game policy as it is affected by the subjects of territory and population and as an ecological factor in the lives of the deer. Antlered game subjected to pursuit by man suffers particular danger of extermination if the pursuit is not very carefully controlled. The mature animals including those with the best-developed antlers are for the most part the ones which come first into rut. The hinds in season at that time are served by these mature stags, and keeping in mind the short period in June during which most of the calves are born, it is apparent that the more mature stags, rutting in a similarly short period in the previous September and October, must be the sires of these calves. There is, however, a tail on the steep frequency curve of calving dates, representing the late calves born in August, September, and even October. These are the progeny of young stags coming late into rut and catching hinds which have come in season a second or third time. Late calves would be better unborn, for the winter takes extra toll of them. Now if too many mature stags are shot in September, breeding is left to the later young stags, and the result is that the extra high winter-mortality factor, working normally on a very small number of late calves, operates on a greater proportion of the calf crop, thus seriously diminishing natural increase and possibly turning the scale in the opposite direction. A population diminishing by such a means is in serious danger of extinction by the many obscure psychological checks to increase which then begin to work. Incidentally, it has often occurred to me how important the secondary sex ratio may be for the survival of small wild stocks. Small samples proverbially give erratic sex ratios, and when this number happens to be the whole population, extinction may be caused by too great a preponderance of one sex to the detriment of the other, especially if the preponderance is of males.

It has been suggested from time to time by those who criticize the private ownership of deer forests that the forest ground of the Scottish Highlands might be pooled under national control or some other form of syndication. Sportsmen would then be allowed by licence to enter the forests at any point and take out one stag or a certain tally. I am not concerned with the human aspect of the politics at the back of the idea, but I am strongly antagonistic towards it from the point of view which interests me most, namely, the preservation of the wild red deer of the Highlands. Human politics cannot be disregarded as an ecological factor in the lives of animals. Forest owners look after the stocks on their ground with care, and the forests are stalked scientifically on the advice of the stalkers, who are employees of the owners. Unfortunately the human race, particularly the mass of the so-called civilized part of it, has shown itself all too often quite unfit collectively to care for a population of wild creatures that have a trophy value. The forest area of Scotland is small (about 2,500,000 acres, not all contiguous) when compared with game areas in the United States or Canada, and institution of a 'one stag law' in Scotland, which could place no territorial responsibility on stalkers or sportsmen, would soon mean the end of the red deer as a wild member of our fauna. In the last paragraph but one I referred to the percentage of cripples. Owners of forests do not suffer bad shots gladly and the code of behaviour is strict. If the forests were open to anybody with a rifle who took out a 'one stag permit' a great deal of damage could be done, and the cripple figures would probably approach the American ones which I have quoted. What is at present clean hunting and a humane method of regulating the numbers of a species which has few natural predators would be degraded to the level of a blood sport.

To return now to the burden of our discussion: numerical treatment of the herds on their territories is necessary, as I have said already, for a clear understanding of density in relation to movement and to the social system. But – I have a rooted dislike of its appearance in black and white because the figures tend to give too much a sense of apple pie order. The census (which again suggests a cut-and dried procedure of exactitude) was made during the course of the work by myself alone in the early months of each year 1934 and 1935, not in one week or any definite period of time. Though I am sufficiently familiar with some of these groups of deer to know the individuals comprising them unmistakably, extreme accuracy is quite impossible to attain and, indeed, it should not be sought on such an area of mainland forest. The figures on paper are apt to suggest too static a condition of the herds, whereas populations of wild animals are always tending towards some form of shift. Figures are wooden, impersonal symbols for the dynamics of population in these highly sentient and elusive creatures. I believe the figures to be close to actuality, however, and the arithmetic to represent the growth and modification of the population as they occur.

There remains to be considered the fertility ratio and the proportion of infant mortality in the population. Some hinds bear a calf each year, others in alternate years, and a hind calves for the first time at four years old. Under favourable conditions a hind may calve at three years, but in the Highlands such an event is uncommon. The average age to which deer may live is approximately 15 years. Cameron (1923) mentions a hind which was well over 20 years old and bred each year. Increase in herds of red deer is not rapid as compared with a flock of Blackface ewes. The fertility of adult hinds is about 60 per cent in the herds I have watched, and it is the same as that given by Cameron (1923) for the island forest of Jura. He gives the mortality of calves to one year old as being 50 per cent. When I first read Cameron's book I thought this calf-mortality rate to be extraordinarily high, and while I did not doubt his figures for Jura I could not think there would be as high a mortality among the calves of mainland herds. Now, however, I have to admit that I have found the same rate of 50 per cent mortality in calves to one year old in the herds under my observation.

TABLE II

Territories and Distribution (calves excluded)

Identification sign	Territory	Class	Numbers
1	Carn na Carnach	Hinds and followers	95
2	Glac Cheann	”	5
3	Pine wood	”	70
4	Polain and Meall Bhuidhe	”	70
5	Gruinard	”	50
6	Beinn Dearg Mhor	”	120
7	Beinn Dearg Bheag	”	80
8	Creag-mheall Medhonach	”	20
9	Creag-mheall Mor	”	60
10	Beinn a' Chaisgein Beag	”	200
11	Caiseamheall and Beinn a' Chaisgein Mor	”	230
A	Glas Thuill	Stags	25
B	Coire Mhuillin	”	20
C	Coire Ghamhna	”	50
D	Gruinard	”	60
E	Beinn Dearg Mhor	”	50
F	Ghiubhsachain	”	30
G	Lower Fisherfield	”	50
H	Beinn a' Chaisgein Beag	”	30 |//

Total 1,315

TABLE III

Type Classes (calves excluded)	Males	Females
Stags, upwards of three years	315	–
Hinds, upwards of three years	–	600
Yearlings	80	80
Two- and three-year-olds	120	120
Totals	515	800
	35.6 per cent	64.4 per cent

The first check to survival of all calves born alive is the activity of predacious animals. Well-grown red deer in Scotland have no carnivorous or raptorial enemies, but the newly-born calves are preyed upon by the fox, the eagle, and the wild cat. When the calves are born in June they lie alone for two to five days before following their mother, and this is the time when the toll is taken. The fox is responsible for most losses of this kind. The vixen has her cubs at the weaning age at the same time, and she finds the calves easy prey if the hinds are absent. I have counted the remains of seven calves at a fox den in the mountains, I have never seen a fox taking a calf, but a stalker told me of a vixen he watched creeping through the heather and keeping out of sight of the hind. Before the fox reached the calf, however, the hind had winded her and came up at the run to strike at the vixen with her forefeet, at which the vixen ran away. Golden eagles take a few deer calves, but not many. There are three or four pairs of these birds on my beat, but I do not think they take half a dozen calves between them in the year. It was in this forest that John Cameron, Shenavall, found an eagle's eyrie with remains of fox cubs! While he was at the eyrie the hen bird returned with another cub in her talons. Such an instance of raptor feeding on carnivore must be rare.

The wild cat takes very few deer calves; only in seasons which are lean of grouse or rabbits might losses be attributed to this animal. A red deer calf has surprising strength and is rather too big for a cat to tackle and kill before the hind would be on the scene.

The next dangerous period for the red deer calf is the bad weather of October. If the hind should become very short of milk or die there is little hope of the calf surviving the winter. Spring takes its toll of all age-classes, but particularly of calves and old animals. I have found many deer newly dead in spring, and they are usually heavily infested with ticks and keds and with lung and intestinal worms. These parasites may be considered to be not so much direct checks on the population as an aggravation of the low physical condition caused by bad weather and lack of herbage. My own estimate of 50 per cent calf mortality in the first year is not based on numbers found dead, but on numbers left alive at the beginning of June.

Townsend and Smith (1933) give figures for white-tail deer from which a calving figure of 57.7 per cent can be deduced. Spiker (1933) in the same bulletin records the heavy winter mortality among the fawns from various causes. Hadwen and Palmer (1922) state that the average calf crop in the reindeer herds of Alaska is 50–60 per cent, though the actual prolificness is nearer 70 per cent. These reindeer are efficiently herded on the ranch system and the annual calf loss in the first winter is estimated at 15 per cent. Klemola (1929), who studied the reindeer breeding industry in Finnish Lapland, says that under good management there is a yearly increase of 30–40 per cent (presumably of the whole herd and not of the cows alone) and the average increase is 25–30 per cent. This figure is reduced considerably in the first winter. We see then that the deer tribe as a whole is subject to a heavy first-year mortality.

Twins are of very rare occurrence in Highland red deer. Sometimes two calves may be seen with one hind, but this is always in the new year and not in that of their birth. The truth is that the mother of one of them has died or been killed and another hind has taken the calf along with her own. Roe deer tell the opposite story. The doe may have twins at the beginning of June, but in the following spring she has only one fawn with her. In north-western Scotland death for one of them is almost inevitable.

There is a general belief that the primary sex ratio of red deer in Scotland is high, and this is linked with another belief that 'eild', 'yeld', or barren hinds conceive a stag calf after their barren year. My own observations support these opinions, but in the latter one I have the evidence afforded by very few barren hinds – less than 20, in fact. Glyn Davies (1931) published a short report on the sex ratio of foetuses from hinds barren in the previous year. Of six 'yeld' hinds shot in 1927, five embryos were male and one female. He examined 17 uteri from 'yeld' hinds in 1928. One contained no embryo, three contained embryos in which sex was not determinable, and the remaining 13 embryos were all male. Miller (1932) followed up these highly suggestive results by a much fuller investigation. He obtained uteri from hinds shot in two seasons – 504 in all. There were 99 non-pregnant uteri (as indicated by an absence of any recognizable corpus luteum in the ovaries as well as absence of the usual uterine signs of pregnancy) and in 134 uteri there were foetuses in which sex differentiation (whether gonadic or external) had not occurred at the time of death. These subtractions left 271 uteri, of which 79 were from milk hinds. From these 51 contained a male foetus – 62.3 per cent; and 28 a female – 37.7 per cent. The 192 uteri from 'yeld' hinds gave 111 male and 81 female foetuses – 57.8 per cent and 42.2 per cent. Miller does not attempt to draw other than tentative conclusions from his figures in view of the relatively large number of foetuses which could not be sexed. At least, his observations do not show a total or very great preponderance of male foetuses conceived by hinds which were barren in the previous year. It would have been interesting to know, however, how many of the 192 'yeld' hinds were pregnant for the first time, for the stag-calf belief does not extend to these.

The primary sex ratio (i.e. the sex ratio at conception) in mammals is always high. The secondary sex ratio (at birth) is lower, but still not equality. In red deer these ratios appear to be rather more wide than in many other mammals, but at the age of one and two years I have been unable to notice any significant difference in the numbers of the sexes and I have entered these age-classes at equality in Table III.

This general belief that 'yeld' hinds throw stag calves has led some forest owners to prohibit their being shot when the few hinds are killed in December and January. Milk hinds are taken instead. Unless the calves are shot as well in these circumstances the practice is a cruel one, for the calf has little chance of surviving without its mother. The toll taken, then, is greater than one, and on probability less than two. The underlying motive, obviously, is to raise the proportion of stags on the forest. If such greed were preceded by a little calculation the futility of the practice would be plain to see. The wise forest owner undertakes the hind shooting partly as a culling operation and partly to get fat barren hinds for the larder. He knows that he must always have more females than males on the forest and, as one man put it to me, a good stock of hinds is better than a deer fence.

NOTE

The figures given on p. 34 for reindeer in Alaska are not sound. The history of that gigantic experiment shows that assessment of range needs never caught up with the true needs. L. J. Palmer raised his acreage figure several times, but the crash in reindeer populations occurred before the true carrying capacity was ascertained. My own notion after visiting most of the quondam reindeer ranges of Alaska in 1952 was that it would be unsafe to stock more intensively than two to the square mile if the lichen supply is to be maintained.

In the course of a census of red deer in Scotland which I am at present directing for the Nature Conservancy, it is apparent there are twice as many hinds as stags, and that despite heavy war-time killings, many forests are carrying a stock of three to 100 acres exclusive of calves of the year. A few areas are carrying five to 100 acres. A constant mistake in Scotland has been the equal toll of stags and hinds. There should be a kill of twice as many hinds as stags, and the overall kill should be a fifth of the whole stock. The hind stock should be kept young, except for a few leaders, and therefore highly productive, and stocking should be below capacity, say one to 60 acres.

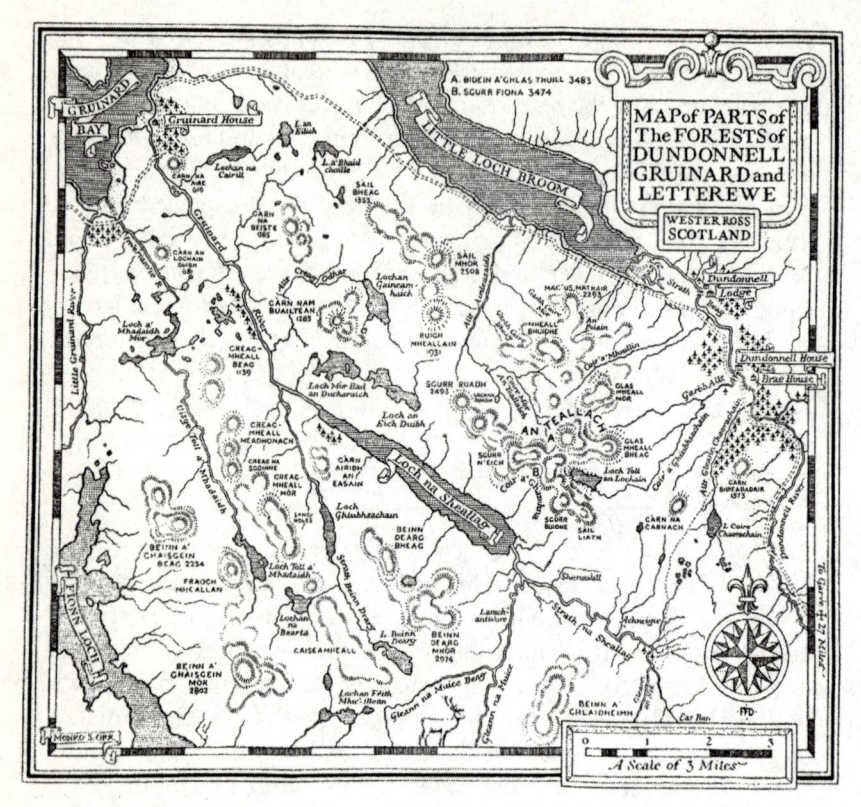

MAP 6

CHAPTER FOUR

Territory and Population (continued)

THE HIND AND STAG territories represented on MAPS 4 and 5 need some individual description and explanation. They should be considered also in relation to the populations represented on them. It has been stated already that the population of deer has been represented on the winter territories, and it is intended to show that stags and hinds are strictly segregated when they are on their winter territories. It will be noted that the summer grazing-grounds are often grazed communally, but that does not mean that the herds of each sex mix together. Once the dispositions of territories on this ground are grasped, the actual social interactions described in the next chapter will be better understood.

The hind territory No. 1 is inhabited by the Carn na Carnach deer. They are most often to be found on the grassy slopes of Carn na Carnach itself, which I have described as being the only part of my ground composed of the undifferentiated Eastern Schist and which is therefore comparatively fertile. That area of this territory which is on the quartzite is but rarely grazed by these hinds, but they are to be seen there more frequently in early spring when the deer go seeking the cotton sedge which grows well in the water-logged conditions it finds in the shallow depressions of the quartzite. The summer ground of these deer is the slopes of the Toll Lochan corrie, and the precipices of Sail Liath and Sgurr Fiona form an effective boundary to their wanderings southwards. For the most part red deer tend to make a glen rather than a ridge their boundary. They like facility for movement over the whole of a hill, which can thus afford them some shelter from whatever quarter the weather may come. The river through a glen is a much better boundary for a deer forest or animal sanctuary than a ridge, and it fits in more accurately with the normal movements of the terrestrial fauna. In this instance, however, the physical character of the ground is sufficient to modify the tendency, and bad weather brings these deer back across Coir' a' Ghiubhsachain on to Carn na Carnach. Coir' a' Ghiubhsachain is frequented by the deer, in the open weather which usually follows the rut and for reasons which I shall attempt to explain in the chapter on movement. The footpath and the burn which run down Gleann Chaorachain on the eastern side of Carn na Carnach are a very sharp boundary for the groups of deer on either side. I have never seen hinds crossing here, but in the rutting season of 1934 I found several stags' tracks in the snow which showed them to have crossed the glen. If these deer are disturbed anywhere on the eastern slopes of the Carn they will make most determined efforts to cross the hill into Coir' a' Ghiubhsachain, even though the disturbance should be from the summits of the Carn, between the deer and the corrie. The sharp quartzite fault below and north of Carn na Carnach, running between Coir' a' Ghiubhsachain and Gleann Chaorachain, forms another strict boundary for the Carn hinds; but if heavy snow is imminent they come through it by the deer path which pierces the fault, and range over the quartzite slabs below and into the drier ground of Glac Cheann. It is only this emergency, however, which brings them off the Carn, and they are back through the fault as soon as conditions allow.

Let us now turn to the stag territory lettered *A*, which is adjacent to the Carn hind territory. I call these stags the Glas Thuill company because they spend a large part of the year either in the Glas Thuill or on the ground just below its lip. It must be very hard weather before they will come into Glac Cheann to stay. Yet when I first came to this work, in March 1934, these stags were in Glac Cheann and on the quartzite slabs below the fault almost the whole of their time, and I was told they had been there all the winter. They were there more frequently in April and May 1935, and every night during the first half of June, but in the winter of 1935–6 there was this reluctance to come into Glac Cheann. There is one reason which would seem sufficient to have brought about this change. The last week of October 1934 was notable for its heavy rain and wild weather which culminated in deep snow on 1 November. The hinds from Carn na Carnach and the stags which were with them at that time came into Glac Cheann and stayed several days. When they returned to their own ground, two hinds with their followers – five beasts in all – stayed in the Glac. They could be seen there most days. At first they were very nervous and were ready to flee from real or imaginary dangers. By Christmas, however, they had settled down in the Glac as their home and they grazed over the quartzite slabs nearby and into Coir' a' Ghiubhsachain. One of these hinds, the leader of the little group of colonists, was of a light sandy colour and easily recognizable. Their movements were easy to watch and they could be seen very often from my house. In summer they joined the Carn na Carnach hinds in the Toll Lochan corrie. They returned to the Glac during the first week in October with a rutting stag, one of the Glas Thuill group, and by the end of November three other stags had been with them in the Glac. But the stags did not stay. Glac Cheann, then, by the winter of 1935–6 had changed from the normal winter retreat of a company of stags to become the well-defined winter territory of a small colonizing group of hinds. The stags were now spending more time in the lower reaches of Coir' a' Ghiubhsachain than they did in the winter of 1933–4. These few hinds will call for attention again later.

From noting this particular instance of change of territory by colonization let us return to the Glas Thuill stags. It will be noticed how, on the eastern corner of their northern boundary, the Garbh Allt has formed a natural barrier. This water could be crossed only with great difficulty on its steep descent from Coir' a' Ghiubhsachain, because it falls through deep clefts of rock. The shape of this stag territory might lead to the question – Why do the stags not graze north of the Garbh Allt? Deer do not choose territory with a natural barrier within it, though frequently such a barrier forms an outer edge to territory. To take an entirely anthropomorphic view, it would be a tactical error to include a natural obstacle in an area where free movement from side to side is desirable, and I think it reasonable to infer that the behaviour of deer is influenced by a similar principle in these circumstances. Large portions of the walls of the Glas Thuill form natural boundaries to movement also, but there is a pass through the corner into Coire Mor which is used by the Glas Thuill stags if they are disturbed or if the wind draws them through on occasion in summer. It is impassable for most of the winter. On the southern side of the territory of these stags we find part of Carn na Carnach to be included, and this is owing to a curious movement made by the stags in late May and early June. The same movement was observed in 1934 and 1935, when the stags would go to the green slopes of Carn na Carnach during the daytime and return to the quartzite slabs and Glac Cheann at night. This occupation of part of the Carn is for a very limited time in the year, when the new antlers are growing and when this ground is growing young grass. The stags keep close together at this time and do not mix with the hinds on the Carn. The route of the stags is along the deer path through the quartzite fault. Some of these stags may be on the Carn again in the rutting season, but as most stags are off their own grounds and on the hind territories at that time, their autumn occupancy is not comparable with this limited May–June movement.

The hind territory No. 2 includes the pine wood at its northeastern end. This wood, it may be remembered from the description in Chapter One, is not the compact mass of trees which the map suggests, but an area of clumps and wide glades. The ground is very wet but the trees afford dry shelter, and the wood is the winter resort of a group of hinds and followers, about 70 in all.

It is often stated that the red deer is a woodland animal by nature and that deforestation and cultivation have driven it to the bare mountain slopes. Broadly this must be true, because there were more deer and more forests in early times, but at this date I do not think the statement needs emphasis. The Dundonnell strath is unusually well covered with natural forest growth and there are no restraining fences, but the woods are not heavily stocked with deer nor do the animals flock into them during bad weather. Where woods are within a territory they are used, but there is apparent no special urge to make for their shelter. There are some stags, usually old and more or less solitary ones, which live in woods at a low altitude and will remain in them throughout the year, but they are not representative of the species as a whole in Scotland.

TERRITORY AND POPULATION (CONTINUED)

Let us compare the red deer with the roe deer, which is a true woodland creature. The pace, if hurried, of the red deer is a long, easy trot, and the animals break into the gallop only when prompted by fright or emergency. Roe deer hardly ever trot: they bound. The spring is from both hind feet, which leave the ground together, and the two forefeet come down together. The animals assume the attitude commonly portrayed in the art of early civilizations. This is an excellent gait for moving over undergrowth and one that an habitual woodland animal might be expected to have developed. Roe deer are extremely difficult to stalk in woods, but if they are out on the open hill, as they are sometimes in summer, they can be approached more easily. The converse is true of red deer if the stalker can rely upon himself to cope with the extra difficulty of seeing the animals in woods before they see him. The red deer appears to have much longer sight than the roe deer, and the roe in its woodlands seems to have a greater capacity for recognizing detail at short distance.

The Coir' a' Mhuillin stags, territory B, are alongside the pinewood hinds, but they do not frequent the wood except in severe weather. In summer they use the pass from Coir' a' Mhuillin into Coire Mor in the same way that the Glas Thuill stags use their pass.

The Polain hinds, territory No. 3, spend their winter in the hollow named An Polain and along the steep slope above the sea loch as far as Ardessie. Heather and good feed are plentiful on this slope in spite of its precipitous character. In summer these hinds cross over the barren dome of Meall Bhuidhe into Coire Mor, and they may be found there until the end of December. The Allt Airdeasaidh forms a natural barrier for them on the north-west.

The next hind territory, No. 4, is of considerable interest. In the first place it is large when compared with the number of hinds on it. Years ago, before sheep were put on the Gruinard ground, there were more hinds on it than there are now, and I am told that there was movement of hinds across the Gruinard River at the ford at the foot of the Allt Loch Ghiubhsachain and where the Allt Creag Odhar enters the river. That movement has stopped now, and there is an encroachment of sheep ground on the present hind territory. Although these few hinds are being encouraged to stay, it would appear that this territory has been a shrinking one for some years; but, as might be expected, the shrinkage in area has lagged behind the fall in numbers. Probably the shrinkage has been checked now, for I have noted a slightly increased number of hinds there during the two years I have quartered the ground. There is another reason why the territory should be comparatively large: namely, it is very much within the stag territory D.

This company of stags and their ground, D, have proved one of the most interesting groups in the forest from the point of view of size and trekking distances. Theirs is the longest territory in the forest, for they winter near the coast round Gruinard Bay and summer in Coire Mor. Before the sheep were put on Gruinard I understand that as many as 40 stags were shot in the season, about half of this number being got during the early part of the season from the rough ground of Carn nam Buailtean and the western slopes of Ruigh Mheallan. The rate of grazing from the winter ground near the coast to the high hills was evidently slow, and it is possible that some of the stags did not go farther than Carn nam Buailtean. Now, however, they do not stay long on the sheep ground but pass right over it into the quietness of Coire Mor, which, incidentally, is over the boundary of the Gruinard Forest. Their movement into Coire Mor is mainly over the broad pass between Ruigh Mheallan and Sgurr Ruadh. The disposition of Loch Mor Bad an Ducharaich and Loch an Eich Duibhe appears to be sufficient check to prevent the Gruinard stags from going forward on to the slopes north-east of Loch na Sheallag. The Gruinard hinds, however, pass regularly over the little burn between the two lochs, and the milk hinds of the group are particularly fond of the rough ground between these lochs and Loch na Sheallag.

The Coir' a' Ghamhna stags, C, have two natural boundaries to their territory – Loch na Sheallag and the ridge of An Teallach. The bog below Shenavall provides excellent wintering and early spring pasture, but if the weather is open the stags are mostly to be found at about the 1,000 feet level above the loch. They are not found much higher on this slope in the winter months because the gullies filled with soft snow are a serious check on lateral movement.

Herbage, too, is scanty on the higher parts of this slope. In summer, however, Coir' a' Ghamhna is a favourite grazing ground of stags, with Coire Mor as an occasional alternative if wind or disturbance provide cause. John Cameron, Shenavall, has tried to make the actual corrie a sanctuary, and I think it held more stags in 1935 than in the previous year. The corrie is so shallow and bare, however, that it has no holding power beyond the short stalking season of August and September.

To conclude our survey of the territories north-east of the Gruinard River and Loch na Sheallag two points are particularly worthy of note. One is the natural barrier which Strath na Sheallag from Loch na Sheallag to the sea forms to general movement. The loch is crossed by rutting stags, but the river even when in small spate can be quite impassable. Secondly, we are impressed with the position of Coire Mor An Teallaich as a summer reservoir of deer. It is difficult of access for man and it is more free of Tabanid flies than all the rest of this ground. It is a fine experience about the middle of July and at the height of the fly season to be sitting on the summit of Sgurr nan Eich, the western buttress of An Teallach, and from that high perch to count the knots of deer in the expanse of Coire Mor. They are easily picked out with the help of the telescope, for the Coire is green, their coats are by this date bright and rufous, and the short alpine herbage does not conceal them. The stags are lying near in below the Sgurr Ruadh ridge and across the head of the corrie. Their new antlers are well grown and look larger than they really are with the covering of velvet. Well below and on the slopes of Meall Bhuidhe are groups of hinds widely spread in their grazing, and the dappled calves are running among them, playing their little games and diving under their dams for a drink. The heat haze is over all and the very blades of grass dance in it. And yet, for all the quiet of this place, I know of no corrie of its size which can be so easily cleared of deer. The utmost care is needed if its peace is to be enjoyed. The stalkers told me this when I started my work, but I had to learn it for myself. A puff of wind not reckoned for will put the deer through the passes into the other high corries, and Coire Mor is empty.

The pine wood, Dundonnell

Now for the ground beyond the river, the south-western half of the total area under discussion. The stag territory *E* slopes down from Beinn Dearg Mhor to the bog at the head of Loch na Sheallag, the bog which is the hub of winter ground for several groups of deer. This territory overlaps much of that of the Beinn Dearg hinds, which are a well-established group on these hills. At the same time it will be noticed that the stag territory *F* lies almost wholly within the Beinn Dearg hind ground No. 6. This is interesting, for the stags established themselves in a very short time during the period I have been working over this area.

It is the usual practice in the management of hill ground, whether deer forest, grouse moor, or sheep farm, to burn part of it each year. The heather in the west Highlands is not strong enough to stand much burning and once in 10 years is quite sufficient. The slopes of Cam Airidh an Easain and the foot of Beinn Dearg Bheag as far up as the head of Loch Ghiubhsachain were burnt very thoroughly in March 1935. In 1934 it was a rare thing to see stags on this stretch, but before the grass began to grow in May 1935 after the burning, a number of stags came in and remained for the growing season. Stags are avid for young grass on burnt ground, that is, just at the time when the new antlers are growing. I have seen them on newly burnt ground pulling at the charred grass. It is not the palatability of the herbage which attracts them in these instances, but the readily assimilable ash which they can obtain and for which they crave. Naturally the new grass on burnt land must contain a higher percentage of mineral ash because the spring rains dissolve some from the burnt grass on the surface of the soil and carry it to the plant roots. Most of these stags came in to the burnt ground from the stag territory G, Lower Fisherfield, and this is certainly the most likely place for them to have come from. I knew these stags intimately, as it happened, but there were a few strangers. There was still a good number of stags left on Lower Fisherfield and new ones came in there during the summer which were strangers to me. But the newcomers were youngsters, mostly three-year-olds, and I am convinced that they came from the Beinn Dearg Bheag hinds and the Beinn a' Chaisgein Beag hinds. There is considerable change in the antlers from year to year in young stags, and as two-year-olds they have not acquired such individuality as would enable me to identify them accurately. Furthermore, it may be easy with practice to recognize a beast on the ground it usually occupies in company with its usual companions, but if, as in the case of a young stag without characteristic antlers, it changes its ground and leaves the maternal group, it may appear to be a complete stranger in its new habitat. The Ghiubhsachain stags, as I have called them, were still in occupation of the ground in the winter of 1935–6, and have provided another example of the influence of man on the territorial affairs of the deer on this ground. The hinds still graze over the ground, and it remains to be seen whether the new stag territory will become permanent.

The Beinn Dearg hinds themselves offer no points of particular interest. Strath Beinn Dearg is good, quiet grazing for them. They are very fond of the hollow – not deep enough to be shown clearly on the map – which is on the western slope of Beinn Dearg Bheag and which is formed by the junction of the gneiss and the Torridonian Sandstone.

The stags in the territory lettered G do not call for much comment. There are some good beasts among them, but the majority of the company is composed of young nondescript animals, and the number of broken and aberrant antlers in this territory is unduly high. The reason is not clear. Over the river in the stag ground H the animals are a particularly good lot, and from them come many of the early rutting stags which range over the hind territories of Beinn a' Chaisgein Beag and Creag-Mheall Mor. There is a distinct tendency among stag companies for age groups to graze together, and this may account for the obvious difference in the animals on the two sides of the Inverianvie River.

The hind territory No. 9 is the home of a large number of big, fine hinds. The size of this territory is not very much greater than that of the Carn na Carnach hinds, No. 1, but there are twice as many deer. There is distinct seasonal movement, however, among the Carn hinds, and for considerable periods some parts of their ground are never grazed. The Toll Lochan corrie, for example, is empty for four or five months in winter and spring, and the quartzite slabs are not grazed during August, September, and October. Seasonal movement is considerable, but general daily movement is slight. Beinn a' Chaisgein Beag does not show the same variety of country, and apart from the summit in times of snow, there is hardly any part of the territory where the hinds may not be some time or other during the winter. Daily treks of a mile or two are common. The principal cotton-sedge areas in the Beinn a' Chaisgein Beag territory are here and there along the course of the Uisge Toll a' Mhadaidh and at the other side of the hill round the foot of the Fionn Loch. It is on these places, between February and May, that the hinds can be expected to be with more likelihood of accuracy than at other times of the winter.

There is an interesting little hind territory on Creag Mheall Meadhonach, No. 8. The hinds were there when I first came, but in no settled fashion. There was one stag with them during the rutting season of 1934, and his isolation from the other breeding grounds was striking. His was the only roar to be heard along that brae face. In 1935 there were two stags with these hinds and others were about. The hinds were now quite settled and were to be seen more about the summit of the hill as if they were extending their territory. I think I must have begun work just too late to witness another example of colonization by a small group of hinds. There can be no doubt that they budded off from the Creag-Mheall Mor group, No. 7.

The territories of Caiseamheail and Creag-Mheall Mor, Nos. 7 and 10, form the main part of the great reservoir of hinds which I mentioned in the previous chapter. I am less familiar with these hinds than with any on the forest and I do not propose, therefore, to discuss them in detail. Their ground, by the comparative uniformity of its vegetative complex, enables them to graze over a great part of it at almost any time of the year, and their seasonal movements consequently are not pronounced.

I have remarked on the discrete nature of the boundaries of winter territory, and it has been possible to check this finding by a feeding experiment with the small group of hinds and followers which colonized Glac Cheann. In January 1935 I began to put down a few handfuls of kibbled maize on a dry patch which was cleared in the middle of the Glac. It was almost a month before the corn was touched by the deer, then only to be scratched about. They were feeding steadily, however, by the middle of February. As the spring advanced and lighter evenings made watching possible, I wanted to know how the deer reacted to my tracks. Handfuls of maize were put down in several places in the Glac and the food had disappeared by the next day. This became something of a game. I would put down a single handful of maize here and there at random, and it was for the deer to find it by the next day. They always won. It was quite easy for them to do so once they had reached the stage of following my track instead of running away from it. But sometimes, both in 1935 and 1936, I would cross the lower end of the Garbh Allt or the Allt Gleann Chaorachain and put handfuls down there. The maize was never eaten, and presumably the Glac deer did not follow my track across their own boundaries. It might be suggested that crossing the water provided sufficient check to prevent the deer following my track farther. I doubt if this could be accepted as sufficient reason, however, because deer willingly cross water in their normal movements, and there were rocks in the beds of the rivers from which they could have taken my scent.

Deer Paths, Wallows and Rubbing Trees

The survey of individual territories which has been made in this chapter would not be complete without mention of deer paths, wallows, and rubbing trees. These may be called public works, but the analogy must not be carried too far. They are not produced by communal effort, but from the repeated usage of individuals. Whilst there appears no reason to infer that these modifications of territory which make for general amenity are developed by the herd for the herd, they provide evidence of the sub-human presence of tradition and they are important objects in the social life of the herds. Red deer have grazed this area in greater or lesser numbers uninterruptedly since prehistoric times, and although we have seen that group territories tend to change in size, shape, and situation by the operation of small factors, it is almost certain that the main deer paths are of immense age. When man occupies a new and roadless country he uses the game trails and the rivers first of all, and these tend to become the main avenues of diffusion and contact until other and more artificial means of communication are established. Game trails and paths are never straight even when the country is sufficiently open and level to allow of their being so; and this is characteristic of many human roads. The main street of Sydney, Australia, is generally supposed to deviate from the straight line because buildings originally sprang up at each side of the trail made by the ox-wagons bringing wool from the interior. Most mammals have this habit of making paths, even to the point of each user putting its feet in the same places along the paths. The semicircular ridges in the mud round a gateway through which cows are accustomed to pass and the diagonal track across a grass field must be familiar to most people. The rabbit-snarer takes advantage of the same habit by setting his wires along the rabbit runs in between the places where the animals normally touch the ground with their feet.

Some of the principal deer paths in the area under discussion have reached the stage of appearing to have been engineered. The passage of countless feet along a steep hillside has pushed stones and soil slightly downhill to make a firm narrow terrace which is the path. Moving deer naturally go slowly and usually in single file. Each herd movement along a path, therefore, makes a good impress. When I first worked over this country I used to sit awhile and consider the best way over each new stretch – to keep out of the bog, not to get stuck in the rocks, not to lose height, to note where the rivers could be crossed. It was a source of delight to find how often I would discover lengths of deer path leading me over the ways I had planned. How unerringly the paths cross gullies and burns at the most convenient places! Deer are nimble creatures and they can leap a five-foot fence without a run if necessity urges, but the fact remains that in their normal life they will make a considerable detour to avoid leaping or difficult climbing. This ground abounds in impassable places and others where there is but one way of getting through, and for any man who is not a mountaineer the deer paths are an infallible guide. Where the deer go man can follow in safety, and he should trust the deer paths rather than his own judgement. The head of Loch Toll an Lochain gives an interesting example. A cliff falls sheer into the lochan at the head and there is a good depth of water there in the driest times. It is not possible to get round the head of the lochan at or near water-level. By going up 200 or 300 feet there is a deer track leading to terraces of grass above the cliff, and by devious ways through the rocks the path shows the only possible route round the head of the lochan without need for actual climbing technique.

There is another example of the fine placing of a deer path on the quartzite slabs to the north of and below Carn na Carnach. The slabs lie at an angle of about 15 degrees from the horizontal and the path crosses them from Glac Cheann to Carn na Carnach, rising from 350 to 600 feet. The slabs are bare in several places and I have remarked already on their slippery nature, of which the deer are well aware. However, the path crosses them in two places where the angle of the slant is reduced to perhaps 10 degrees for the width of a foot or thereabouts before continuing the general slope down to the burn. The help which this affords in crossing the slabs is substantial, and it is enhanced by slight chips in the quartzite here and there which are used by the deer as footholds. If MAP 7 is studied in relation to the map and the geology of the country, it will be apparent that the paths on the Torridonian Sandstone are longer than those on the Lewisian Gneiss. This is understandable when the broken nature of the gneiss is remembered and the less marked seasonal movements thereon. There are often slight breaks in the paths; some are only a few yards long, and the black ink of the map makes their representation much more prominent than the paths appear in actuality. The most obvious paths are those on the upper slopes of the Torridonian hills where there is no peat. Those on the Sgurr Ruadh side of Coire Mor, in the Glas Thuill, and round the southern edge of Beinn Dearg Mhor are very good ones and would persist for a long time even if the deer disappeared. The tactical value and structural quality of these paths are facts worthy of repetition and emphasis, and at first sight it is difficult to imagine that they have been produced by individuals seeking and following the easiest way, thereafter using it by memory, and that they are not preconceived works.

Wallows are in the same category as deer paths. A wallow is a place in the peat bog, scratched by the forefeet of the deer where the peat can be churned up into a creamy consistency by the animals' feet and which will give sufficient depth in which to roll. Red deer wallow in May and early June, and again in September and October. A large, well-established wallow may be as wide as five yards across. This type has usually a central hard core which is not scratched into the wallow. One which is illustrated on p138 has a root of bogwood as centrepoint. Some wallows have room for one beast only at a time but may be used by several deer in the course of the day. During the periods when the wallows are not used they begin to assume a similar appearance to the surrounding bog, but as soon as it is May or September the deer break them open in most determined fashion.

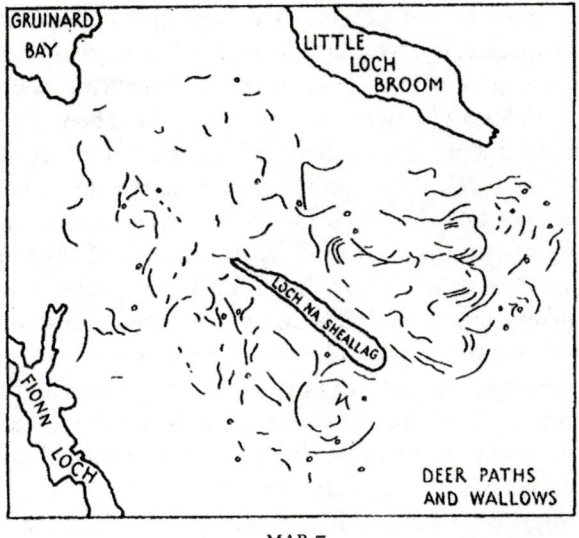

MAP 7

Stags and hinds make different kinds of wallows, and I can state only the bare objective fact without offering reasons. It is the stag wallow in which the peat is worked up into a medium creamy consistency. The stags wallow more inveterately than the hinds, and to a much greater extent in autumn. The hinds roll in clear water but not in supersaturated peat. Their peat wallow is much drier and is almost crumbly in character, providing a place in which to roll rather than to wallow. One of these, below a peat hag, which is a characteristic position, is illustrated on p138. Peat, unfortunately, is not a good substance to photograph, and although a good day and the best time were chosen, the dark granular peat has not allowed a very good result. The possible functions of wallows will be discussed later in the chapter on reproduction under the heading of the sexual psycho-physiology of the stag; they are mentioned here because of their territorial significance as objects of common usage by the members of the herds.

Rubbing trees occur only in those territories where there are coniferous trees, preferably pine trees. Birches and alders do not appear to be used, even where they occur singly in positions which might be expected to attract the animals. A pine tree has a rough bark which may provide a good counter-irritant scratch, but before long the bark has quite worn away and the wood beneath becomes highly polished. One of these trees is illustrated on p 156. The trees continue to be used in their polished state, so that the roughness of the bark cannot be held to provide the attraction. Perhaps the odorous quality of coniferous wood has an insecticidal value, and certainly when one of these trees has been killed a new one is used. Large numbers of deer have no chance of rubbing against a pine tree. In one forest I know, a stob of larch or pine is driven into the ground far up on the stalking path, and the ponies are tethered to this in the stalking season. The stob, however, attracts the deer as a rubbing post and it is soon worn through; even a wrapping of barbed wire provides only a temporary deterrent.

It is amusing to watch a hind having a rub. She starts with her fore face, first one side and then the other, not too hard. Then the forehead with more vigour, and perhaps the eyes are closed. The back of the ears is a tricky place, calling for finesse, and the eyes are very wide open. The neck provides an ecstasy of rubbing and she rubs as if she would push the tree over. Then a shake before going on to the ribs. She presses against the tree, and her feet wedged in the ground keep her tight up to it. To finish this apparently invigorating toilet she rubs her hind quarters to and fro on the tree, and her face assumes that foolish look, of half-closed eyes and dropped lower lip, which is characteristic of horses when they are rubbing their buttocks against the paddock rails. Then she has a shake and away she goes, head up, ears forward, and with a fine springy step. Stags rub their antlers against the rubbing trees, but not usually at the time when they are shedding the velvet because they are then high on the hills away from the midges. However, in eastern Scottish forests, where there are more pine trees and less midges, the stags do use trees for gently rubbing away the velvet. Stags also use their hind feet for this purpose. On the gneiss territories where there are no trees I have not seen the deer using particular rocks for rubbing, nor have I seen them going through the same purposeful and complete rub down as those which have pine trees for the purpose. Rubbing trees, then, are not invariable and indispensable components of territories, as are paths and wallows, but they are established wherever coniferous trees occur.

NOTE

I have recorded in this chapter two examples of colonization of former stag territory by small groups of hinds, and as the whole of the ground over which I worked was populated by deer, I was not able to give examples of colonization of new ground. Dr Peter Delap, who has recorded in *The Irish Naturalist's Journal*, July 1936, his observations on the spread of red deer in the Wicklow mountains, has kindly written to me and pointed to the fact that he found stags to be the initial colonists where the species was increasing its range, and remarked that both my examples of colonization were of hinds. His timely letter enables me to correct any error of emphasis I may have made by citing only my limited observations without further comment. Dr Delap's findings are parallel with what may be observed in Perthshire and Aberdeenshire, where the Highland deer forests end and the sheep farms begin. The deer are attempting to colonize there and the forerunners are always stags. My own examples show the next stage, of hinds colonizing former stag territory. As I have said on p74, the stags are much more given to wandering than hinds, and it may be accepted as typical that when the species colonizes new ground, groups of males are the first to enter it.

CHAPTER FIVE

Territory and the Social System

SOCIALITY AND TERRITORY are related concepts. Territory, even, may be held to be embraced within the idea of sociality. It was necessary to touch on the social system of red deer at the beginning of the chapters on territory, and the final consideration of territory can well be fitted into this present discussion on cervine sociality. The whole of this book bears on sociality, but here the system in this species must be given a definite shape.

What are the advantages of social habits to the higher animals? Thomson and Geddes in their last great sociobiological work (1931) answered this question at considerable length. These headings can be summarized from their discussion of the value of sociality:

1. Strength of union. The whole is greater than the sum of its parts.
2. Co-operation, making for efficiency.
3. Possible permanent products and educative potency of the society.
4. Division of labour.
5. Sociality fosters the evolution of intelligence.
6. Social habit works in effect towards a moral and ethical standard. (This aspect of animal sociality has received considerable space in Dr Albert Schweitzer's Gifford Lectures, 1935.)
7. Sociality allows for the trial of variations with a freedom not possible in solitary animals.

There is also a physiological advantage in an association of animals at an optimum density which as yet we are unable to explain adequately. Allee (1932) has brought together some data on this subject, e.g. a number of tadpoles regenerate lopped tails sooner than single ones; fruit flies (*Drosophila*) grow larger and live longer in a medium than in high or low densities. As this question stands at present, the physiological value of association is alone apparent; the relation to sociality is obscure and may be non-existent.

Red deer are essentially gregarious and their social system is matriarchal. Landseer showed himself a faulty observer when he entitled his picture 'Monarch of the Glen', yet this phrase has persisted in the popular imagination. The stag never attains to leadership, and the first stag in rut which may be running round a group of 50 hinds still has no power of leadership. When more stags come into rut and the first one loses some of his hinds, he is not master of the whole territory of hinds and the stags which are then with them, but only of the few he can keep together. The sexes are separate for the greater part of the year, and the social constitution of hind groups and stag companies presents many points of difference and contrast.

The Hind Group

The outstanding feature of the hind group is its cohesion which, doubtless, is derived from the stability of the family. Maternal care is protracted in the red deer, extending to the third year of life of the offspring. Thus each hind may have two or three followers, and some of the other adult hinds may be the earlier offspring of a hind still in the group. The protracted education and the high degree of sociality in this species are no doubt indissolubly linked. One hind, a mature and often an old one, is the leader of each group of hinds and followers, and usually she has a calf at foot. I have never seen an instance of her supremacy being challenged, and it is probable that she is not superseded until death removes her. I have said that usually she has a calf with her, from which we may infer that the leading hind of a group is a regular breeder and that that part of her maternal emotions which embraces the herd as well as her own offspring is stimulated by the birth of each year's calf. In fact, a hind which ceases to be a regular breeder soon ceases to be leader.

The important role of the anterior lobe of the pituitary body (the endocrine organ situated at the base of the brain) in reproductive processes and lactation is well known. Wiesner and Sheard (1933) have demonstrated the significance of the anterior pituitary hormone in initiating and maintaining maternal behaviour in the rat, and it is reasonable to believe that the same mechanism is operating in the hind. Lactation, which is initiated by anterior pituitary hormone – Turner (1934) – extends over the period of a whole year, a fact to be noted particularly when we remember the heavy first-year mortality of deer and the slight chance of survival of a calf losing its dam in the winter. Turner shows that in addition to actual secretion being brought about by anterior pituitary hormone, galactin or prolactin, the continuous presence of the hormone is necessary for persistence of milk production. The inference is that high and prolonged activity of the anterior pituitary body in the reproductive processes of the red deer is a factor of prime importance concerning the sociality of the species. Indeed, however neat and plausible the advantages of sociality to a species may appear as teleological conclusions, we are always brought up against the fact that sociality has its roots in the physiological and psychological processes of reproduction.

Below the leader there is the usual well-defined order of precedence to be noticed in groups of the higher animals. The deer are not quarrelsome as a group, but if two hinds do fall out they may take sly slaps at each other for most of a day with their able forefeet. The leader is assisted by a second hind who takes the rear position in any considerable movement of the herd. This unquestioned leadership is of the mother type and bears no relation to the masculine egocentric kind which enjoys power for power's sake. The leading hind is constantly anxious for the herd welfare, and the epithet she has earned from the stalker of 'herself of the long neck' aptly describes the constant raising of her head to smell, hear, and see. Other hinds will help her in this anxious task and take their turns as smellers-hearers-watchers, but the leader is always the most alert of the group. The human observer of red deer must make certain of the leader's identity before he undertakes a stalk in order to get nearer to them, and her movements must be closely watched all the time.

The females of a group are much more alert and inquisitive than the young males. As I have said, the leading hind takes most responsibility, and once she is noted, it is surprising what chances an approaching observer may take with the staggies in the group. As calves they seem as sharp as their sisters, but as yearlings and up to three years old, when they leave the hind group, they appear to take no part in the general watchfulness. This may be looked upon as a sexual difference and as a concomitant of male sexual immaturity. Alertness, individually and on behalf of the group, obviously increases and improves with age. This means, of course, that discrimination also improves. It is possible by crawling to approach a group of deer with a little staggie watching you. His attention wavers from time to time and he looks away. His interest, though returning, is not backed by discrimination, and he does not bother to rise or warn the herd of this human thing unknown on the horizontal plane. But you cannot fool an adult hind like that. Once a young stag has left the hind group he becomes watchful and attentive to detail. At the actual time of leaving he may or may not be sexually mature, but his social environment is altered and his behaviour changes to become adapted to the new conditions in which he finds himself as a member of a stag company, a looser association more dependent on the rule of each for himself.

One of the most delightful examples of social behaviour to watch in red deer is the movement of a hind group from a source of disturbance (let us suppose it to be the observer), which entails the herd's disappearing for a short time when crossing a depression in the ground. The hind at the rear stops at the last point in sight of the observer and turns about. She stands motionless, watching, until the leader and the bulk of the herd begin to appear again beyond the depression. Then the leader and the herd stop and face the disturbance, and the rear-guard hind trots over the hollow out of sight and rejoins her fellows. I do not want to suggest that the hind in the rear is able to assess exactly when the leader and the main body of the group come into view of the observer again. It is probable that the halt of the leading hind in the position where she can see the observer is the signal which prompts the retreat of the hind in the rear. As I shall have cause to explain later, deer have a marked objection to allowing any person or object out of their sight which they may think to be a source of danger. Even when the herd has to pass out of sight the last hind will sometimes wait for several minutes at the point of disappearance.

One day in the early spring of 1935 when the snow had been low I was lying in the rocks above and east of Achneigie. There was a bunch of hinds just above Achneigie and I saw them come up at a trot on to the broad shelf between the house and me. Then I saw the stalker's wife going for her cows with the dog. When the leading hind had brought the herd to a point just below me and out of sight of the floor of the strath she turned about, left the herd, and trotted back to near the edge of the shelf. I imagine that from the strath only her head could have been seen. There she stood and watched the whole operation of the cows being brought in, and not until cows, dog, and woman had disappeared into the byre did the hind walk back the 300 yards to the rest of her group. To note the contrast between the normal orderly movement of the herd and a movement in which leadership has gone, it is only necessary to see a stampede – a rare occurrence which will be described in the next chapter.

Here is another example of the watchfulness of the leading hind. The few deer in Glac Cheann which I fed during the winters of 1934–5 and 1935–6 never grew tame enough to approach the food while I was near. In the worst of the weather they would stand away at a distance of 50 to 100 yards, there to wait until I left the Glac. They were often on the edge of the quartzite slabs, 300 or 400 yards away and just out of sight of the actual feeding-place. I hid in the heather behind a knoll one evening in February 1936, 10 yards from the food. The five deer were a long time in coming down and the leader was not quite easy. As she usually watched me leave the Glac before coming down to the food it is possible that the pattern was incomplete for her on this occasion. She stopped five yards from the food, her muzzle raised, and the others ran down to the maize and began eating. She looked about and walked round with stilted steps. Two minutes had passed and the corn was fast disappearing. She moved five yards up the hill beyond the feeding-place and from there she could see the top of my head. She looked at me, moved her head this way and that, and walked a few steps higher to see all of me. The others were still feeding and walking within a few feet of me. The leader barked once. The four swung up their heads, ran up to her, and turned to look in my direction. They all stood still 25 yards away. The hind barked again and I sat up in full view. They did not move. I got up and walked away rather shamefacedly before their gaze. They knew me well enough not to take precipitate flight. When I was out of the Glac and across the glen at Brae House I looked through the glass at them still standing there. They could, of course, see me walk up to the house. Then I saw the leader trot down first to the remains of the food and the others followed her. By relating this incident I do not wish to imply conscious altruism on the part of the leader, but in her alone of the group did watchfulness override the food-drive. Her behaviour was not egocentric.

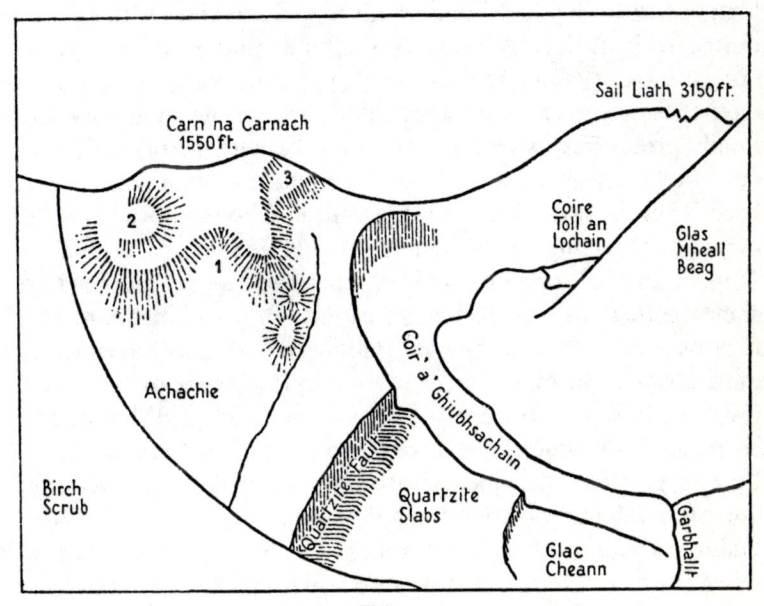

FIG 3
Diagram of northern portion of Carn na Carnach Hind Territory.

In the previous chapters we have seen that there are hind territories carrying varying numbers of hinds from five to over 200. When the whole group is together orderliness is most apparent, but the sociality of the hinds is not so simple as that. The whole number of hinds on a territory is one group, though they are not usually all together. Various factors, to be discussed later, bring the total number into one herd, but the hinds break up into smaller groups and families and there are good reasons for certain types of reassortment in the course of the year. It would be tedious to go through each territory noting the changes in social grouping: we will consider one only. Let us imagine the Carn na Carnach hind territory, with the help of the diagrammatic sketch-map, FIG. 3. The total of 95 deer on the Carn includes between 40 and 45 adult hinds. These are divided into three main families, and their favourite grazing- and resting-places are marked 1, 2, and 3 on the sketch-map. Under conditions of good weather and in daylight, at times of little herd movement, these three family groups may split up still further into individual families, so that the whole of the Carn is dotted with its deer. Any one of the hinds may wander anywhere on the territory, and as a group they do so, but as families they have these preferences for particular parts of their territory. Each family group has its leader, but when the herd is together as one unit the family group leaders submit to the one leading hind of the herd. Their behaviour is modified to fit in with the normal changes of social grouping. In November and December, occasionally in January and February, and in June when the deer are moving to the summer grazings, other changes of grouping within the territory may be apparent, and they are mainly referable to the exigencies of maternity. The milk hinds and their followers are to be found on Carn na Carnach for most of the winter. The 'yeld' hinds, more correctly those which have no calf at foot, may be found in the Toll Lochan corrie in the open weather of November and December, and often on a good day later in the winter they will be in Coir' a' Ghiubhsachain. When June comes, the truly barren hinds will move up first. There is a distinct tendency in late autumn for the three-year-old hinds to form close little bunches of their own, and they will stay up in the Toll Lochan corrie as late in the back end of the year as the weather will allow them. It is noticeable that when these small companies of young hinds are disturbed they move off in much less orderly fashion than a complete hind group. They are much more mobile and do not stand motionless for the same length of time after they have seen an intruder.

Calving time is from about 7 June to 15 July in Highland red deer, and most of the calves are dropped between 15 and 25 June. There is a sharp falling off in the numbers calving early in July, and only occasional calves are born in August and September. One hind on Carn na Carnach calved in the beginning of October 1934. The calf lived through the mild winter and went to the tops with her dam in June 1935. At calving time the group breaks up and each pregnant hind goes on her own. Her followers are with her, but when the calf is born she drives off the youngsters. This has little effect on the yearling, which treats the chasings as extra play. The two- and three-year-olds are content to be driven off to a distance. It is at this time that the young stags leave the hind groups, move off the ground, and join companies of older stags. Six or eight young stags left the Carn na Carnach group in June 1935 in this way. Their maternal group is broken up, there are no large bands of hinds to mix with, and their own mothers are actively driving them away. The little company of three-year-olds – and occasionally two-year-olds – goes away on its own, and in 1935 they joined the Coir' a' Ghamhna stags. It is at this time that the young stags begin to look round for themselves, but they take time to learn their individual responsibility. The young stags leaving a hind group flock together, and this probably is the basis of the observation that age-classes of stags tend to keep together.

The hinds drop their calves in the heather, preferably rather long heather, and in the sheltered portions of their upper winter territories, though there does not appear to be any effort to hide the calves in particularly secluded places. The calf lies alone for two to five days and it is fed by the mother twice a day. During this time the calf will change its position several times, leaving soiled ground for fresh two or three yards away, turning round several times in the new place to make a bed. Once the calf is strong on the legs it follows its mother very closely, and within a few days the hind, the calf, and possibly the yearling and two-year-old move off to the high ground. There they join the barren hinds and followers which have preceded them. By this time the yearling is no longer being driven away from the calf by the dam. A calf sucks its mother right through the winter and late into the spring, and it is not at all unusual to see both the yearling and the calf of the year taking milk from the same hind. Throughout the summer the hind groups are well knit and they move about the large grazing-faces of the high hills in full complement.

We see, then, that in the course of the year there is constant rearrangement of grouping within the main hind group, but it is no random affair. There is a good reason for every change, and the point I want to emphasize is that the hind group still remains one 'city' even when sub-groups of its members may be two miles apart. Orderliness is apparent throughout, and when the whole group assembles, from time to time, family discipline and leadership give way to that exercised by the leading hind of the territory. Individual and particular examples of behaviour within the hind group will be described in later chapters. We have confined ourselves here for the most part to general trends.

Stag Companies

Stags form companies on their territories of a much looser character than the hind groups. The characteristic differences between stag companies and hind groups, socially and territorially, are essentially extensions of primary sexual physiological and psychological differences. As Fleape has written in his *Sex Antagonism* (1913):

> The Male and the Female individual may be compared in various ways with the spermatozoon and ovum. The Male is active and roaming, he hunts for his partner and is an expender of energy; the Female is passive, sedentary, one who waits for her partner, and is a conserver of energy.

Thomson and Geddes (1931) have tabulated the average differences between male and female, and those particularly applicable to the social and territorial behaviour of red deer are quoted below:

Male	Female
Rising to more intense outbursts of energy.	Capable of more patient endurance.
More impetuous and experimental.	More persistent and conservative.
More divergent from the youthful type.	Nearer the youthful type.
Often more variable.	Often less variable.
Making more of sex gratification.	Making more of the family.

The thesis of Geddes and Thomson (1890) that feminine metabolism is anabolic and integrative and masculine metabolism katabolic or destructive still holds in its broad sense, and, as the quotation from Heape suggests, the behaviour developing in each sex is strictly comparable. The basic difference is endocrine in origin, no doubt, and we shall have to touch on it in a later chapter.

The stag company is a number of egocentric males and is a very loose organization. There is no apparent leader, though one animal may be in a position to bully the rest, which is quite a different thing. If a stag company is disturbed the animals may run away in single file or they may split up and scatter, but the single file is not a led retreat. The stag farthest away from the source of disturbance takes front place whether he is just a young staggie or a mature beast. The formation of single file is an interesting one to the animal psychologist. Sometimes it may be the most efficient mode of progression for a number of animals, but this explanation should not be accepted too readily. There is no doubt that there is a tendency in many forms of animal life to follow a line or a single file. The horse needs no guidance in the furrow, Köhler's apes played follow-my-leader, children playing naturally do so, and in human life generally the tendency is all too common. I believe the origins are the same throughout.

Cameron (1923) relates an amusing incident which illustrates to what length the disposition to single file may be carried. Some stags were grazing round the skeleton of a beast and one of them was rubbing his antlers among the bones. This stag threw up his head when Cameron disturbed the band, and brought up a festoon of rattling bones on his antlers. 'Bones' got second place in the rush that followed, and the noise of the rattling bones made the first stag go faster and farther. The faster he went, the faster went 'Bones' and the others behind him, and the more noise there was altogether. The stags went far out of sight. I have noticed repeatedly how a number of stags will frighten each other into movement from a source of disturbance which would not have moved them individually or in twos and threes. But one runs a few steps, the others begin to move, the first one jerks again in consequence of the movement round him, and a general movement is started.

Stags are much more given to wandering on their territory than hinds, and, as far as I have been able to see, not always with obvious reasons which can be adduced to account for the movements of the hinds. Presumably they move because they wish to (and very often this is a better reason than has been allowed), and disturbance moves them or splits up the company more readily than it does the hind groups. Stags have intimate knowledge of a wide tract of country beyond their normal territory. This is understandable when it is realized that they leave it normally at rutting-time and may travel many miles. But the looseness of the company may be shown by the following examples of change of companies between winter and summer. One stag which wintered on Beinn Dearg Mhor with several others was well known to me. He was a 'switch' except for one small shoulder point on the antlers, and the tips were very close. This beast disappeared in June 1934 and early in July I found him in Coir' a' Ghamhna. By the end of October he was back on his old ground behind Larachantivore with his winter friends. He left them again in July 1935 and I saw him on An Teallach soon after. Another stag with backcurving antlers which wintered normally at Inverbroom, 15 miles away, was seen in Glen Muic Beag by Donald MacDonald, Larachantivore, in August 1934. This type of movement will call for further mention in the next chapter.

As the velvet is shed from the antlers in August, the stags appear to gather together more closely in companies, and there is daily movement which may involve the whole length of their summer territories. But as the rut approaches, what previously has been play and high spirits becomes serious quarrelsomeness, and the companies are broken up by each newly rutting stag trotting away from his fellows.

The stag companies frequently split into their approximate age-classes when grazing and resting, and this presumably is a social tendency. At the same time it is quite common in summer to see an old stag with one or two young ones keeping persistently away from the rest. This has been noted before, by Millais (1906) and Cameron (1923), and the usual explanation is that the young staggies do the watching for the old one, which can then take his ease the better. Here is an incident I watched high up on the southern face of Sgurr Ruadh about noon early in July 1934. I was 400 yards away and saw all this through binoculars. A prime stag of 10 points showing through his velvet was lying down, his legs under him, head forward and eyes closed. A small staggie was lying 10 yards away, chewing his cud and turning his head every minute or so to look about him. He sees me. Cudding stops. Five seconds pass and he does not move. Then he is on his feet and trotting to the big stag. The staggie lowers his muzzle towards the big fellow, coming within a yard of him. They both lift their heads together and look intently in my direction. Why does the big stag look so certainly towards me, for the staggie is behind him? Five minutes pass; I do not move; they do not. The sun is crosswise to us, if anything in my favour. The big stag lowers his head again, the staggie lowers his and walks away a few yards, not quite to the same place as he was before. He looks towards me, lies down again, and in another minute goes on chewing his cud. The big stag has his eyes closed again and the heat haze dances over everything. After 10 minutes I move on to my knees. The staggie sees me and runs to his big friend. The stag gets on his feet this time and looks unerringly in my direction. But the air is dancing, the wind is in my favour, and the sun a little more so than it was last time he looked, and I am very still. Five minutes more of intent gazing; they lower their heads together, graze perfunctorily for a few yards, and lie down again. I creep away and leave them in peace.

Sometimes an old stag will become anti-social and, either alone or with a fellow of like inclinations, will keep low all through the summer. All day he lies in the bracken which shelters him in some measure from the flies, and he raids the cornfields in the strath at night. These solitary stags frequently fall to a waiting rifle in the moonlight, and it is usual to find that the beast's incisor teeth have disappeared. The solitary or semi-solitary old male is a common figure in many of the higher animal communities from the elephant and the cattle kind to man. In fact, he is a social phenomenon. What is the solitary stag's daily life? What does he do? Nothing, for the most part. He does not go to the hinds at rutting-time, his antlers are poor, and the gonads do not show seasonal enlargement. Unfortunately, during the period of my work I have not had the opportunity of dissecting such a beast. His movements are little more than those between his feeding-place and his sleeping-place, which latter becomes bare and hard from long usage. He eats and sleeps and looks morose.

It is a noticeable fact, and one often referred to in natural history and sporting books, that when both groups of hinds and companies of stags are resting, the individuals are disposed in such a manner that every point of the compass is covered by one or other of them. Single deer tend to lie back to the wind, facing downhill, and in this way proximity of danger is carried by scent from behind and vision is responsible for guarding the expanse below. To repeat, the resting herd is variously disposed, and the result has a value for the well-being of the herd, but I think the element of chance should be given more weight instead of looking for an *ad hoc* cause for the effect. I have watched for long periods many knots of resting deer, and it is usual for most of the animals to change their lying positions at least

Beinn Dearg Mhor, 2,974 feet. (Torridonian Sandstone)
Strath na Sheallag, not visible, is below

once during a three- or four-hour rest, and an horizon is sometimes left uncovered for a time. What gives a much more definite inkling of social behaviour is the fact that the leading hind takes the position at resting-time from which most can be seen of the country and the herd. The stags do not show this delegation of responsibility so markedly, for each keeps a good look-out for himself.

Wounded deer become solitary until they die or are healed. Stags suffer most in this way, but I have seen lame hinds as well. The effect on the herd of this retreat from society by wounded animals must be for its good, though once again I state merely an objective fact and its result without drawing a teleological conclusion. The flesh of deer, like that of many herbivorous animals, seems highly immune from general sepsis. The deer which suffers a broken leg, probably a compound fracture if caused by a bullet, goes very thin and looks miserable until the leg sloughs off. Then the animal regains its condition in a very short time and rejoins the herd.

There is a hind on Carn na Carnach with only three legs, one of the forelegs having gone from high at the shoulder. The sound foreleg has come into a more central position and the hind gets about as well as her fellows. She was useful to me in that I was able to follow the movements of her group with considerable accuracy when I began work two years ago. Her bobbing gait allowed identification of her group from a long distance.

The Harem

There remains to be considered the social structure and significance of the harem, a unit which has been in the past frequently misinterpreted. We are constantly warned in observing animal behaviour to avoid the anthropomorphic point of view. Carried far enough, obedience to this injunction brings an observer to the point of not being able to see the wood for the trees. The harem of red deer, however, is an excellent example of the danger of being unconsciously influenced by the standpoint of the human patriarchal society. The sociality of red deer is matriarchal, and the apparent modification of this structure for a short season of the year does not break up the matriarchy and establish a patriarchy. The word patriarchy implies governance by the male of the family. Now however eager and active the stag may be he does not gather round him (as I have sometimes heard it said) a number of hinds. His action is similar to that of a collie keeping together a knot of sheep. The hinds are impassive to his activity. His interest in them is purely concerned with sexual gratification, and his activity is equally divided between keeping his hinds together and keeping other stags away. This interest and activity absorb so much of his attention that he fails to look about for disturbance or possible danger. The hinds are as they always are: the leader is watchful and all do their share. The egocentric nature of the stag's apparent dominance of the group is shown when disturbance becomes an actuality. The leading hind barks and the group gathers behind her in retreat. The stag may go with the group, not at the head of it or bringing up the rear, but with no responsibility at all. Or, what is equally likely, he is away on his own road, careless of his harem and his lust forgotten till he is out of danger. We must realize then, the dual nature of the harem – the continuance of normal care and responsibility for the group by the leading hind, and the more spectacular but extremely circumscribed, egocentric and casual dominance of the stag.

NOTE

The deer forests of the Highlands have always tended to become fields for political contention, and though that contention has sharpened and taken a shift towards questioning stocking capacity and off-take, there has also been a more general realization that deer are important and necessary members of the grazing complex of the wilder Highlands. I have added a note on p45 on stocking capacity and have criticized the all too common policy of killing too few deer. That criticism is on the ground of humanity and of common-sense management. We have no large predator on deer in Scotland (see Note on p149) and man must fill the place of the wolf conscientiously. Where large predators exist, the deer stock is found to be young, and therefore highly productive. A hill sheep stock in Scotland is kept young by the farmer rigidly culling after crops of lambs. The deer forests of Scotland will be much more productive if the hind stock is kept young. But at this point I would counsel extreme care because of the highly developed social system of red deer. The leading hinds carry the tradition of what might be called groundsmanship and their skill and knowledge must be maintained in the stock if the forest is to run efficiently. A constant succession of leading hinds must be kept going. Their influence in the forest is not only for the welfare of the deer, for where a small hill sheep stock is kept, as is so frequent now, the groundsmanship of the leading hinds has a value for the sheep; for example, in heavy snow. The sheep watch the deer and use the paths the deer make in the snow between the feeding areas blown clear. The leading hinds decide these paths. This help of deer to sheep was beautifully apparent in the winter of 1954-5.

CHAPTER SIX

Some Social Factors and Comparisons

The Voice

VOICE IS A SOCIAL ASSET, and, however restricted its range in some animals may sound to our ears, it plays an important part in animal sociality. Some animals, such as dogs, horses, sheep, cattle, and roe deer, have vocal powers at all times of the year, and apart from a certain greater strength and timbre in the male sex, the voices of male and female are similar. The voices of our domesticated animals have degrees of expression which a constant attendant is well able to understand and from which the desires of the animal can be accurately inferred.

Let us digress for a moment to consider the voice of the cow, one of man's most delightful and expressive companions. Traditionally she moos, but there is the tiny expectant moo of the cow who has finished her turnips and is ready for some hay, and there is the greedy moo of the cow who sees the skep of corn-lick coming along. The monotonous and loud bawl of hungry cattle is unpleasant to hear. There is a distorted moo which pain evokes, the high-pitched blare of the cow in heat which carries for a mile, and the low muffled moo of anger and frustration. Maternity calls forth a small, short, anxious moo, often repeated. This range of sounds, all classed under 'moo', covers a wide range of emotions. The nuances of expression which man comes to understand are immediately comprehensible to other cattle and are capable of arousing similar emotions in them and consequent herd behaviour.

The voice of the red deer has marked sexual differences not only in type and volume but in the times and circumstances in which the voice is used. The hind is able to make a sound like a sharp staccato bark, one at a time, at intervals of five to 15 seconds. This sound is made only when there is a source of disturbance to the herd, and it is made by the leader usually or by the hind who discovered the trouble. Immediately the whole herd is alert, but not necessarily frightened. Thereafter, maybe not a beast moving, only the leader continues to bark at intervals. If they move off, it will be silently, and the hind may bark each time the group stops to look back. If the group is just a family of three, only the mother will bark. Never does the whole group of females indulge in barking. There is in this species, in fact, a social discipline in the use of oral sound among the females! The bark of the hind carries far, and all deer within hearing take notice of it whether they are of her group or sex or not. In barking, a hind opens her mouth momentarily and gives a slight chuck of the head. The stag is able to give tongue only during rutting season and for a short period immediately after. The sound is a roar, often repeated. He stands squarely on his four legs, extends his neck and head, and sound rolls forth from a wide-open mouth shaped like an O. This roar is impressive and doubtless that is what is intended to be. It is the embodiment of lust and its derivative emotions of challenge, anger, and jealousy. The sound at night time in a glen full of rutting stags has a curious quality for the human listener. It is brave enough, but there is a sadness and inarticulateness in it somewhere, a touch of frustration. By November the sound has degenerated to a low moan, infrequently given. For the rest of the year the stag is silent. The stag does not bark except when immature, and I have heard it happen on three occasions only, each time when a young staggie has been alone and I have surprised him – once in Gruinard in June, once in Fisherfield in August, and once on Carn na Carnach in December. These times, except perhaps the last, do not coincide with a possible development of the sexual voice. Once maturity is reached the rare male bark is heard no more.

    I have remarked on the great volume of sound of the stag's roar, so different from the sharp, incisive bark of the hind. For all its volume the roar does not carry far. The wind plays queer tricks, but I have often been within 300 yards of a roaring stag and never heard a sound. A hind's bark would come through this same wind. After all, the bark of warning is needed afar; the sexual roar of challenge is a sound for short distances.

Deer calves have their own small voices, and the hinds will make intimate grunts which are meant for the calves alone and not for the herd. If you are near enough to a herd of hinds and calves in July these family sounds are pleasant and almost continuous.

Play

The tendency to play which is common among the higher animals, especially among those which have a well developed sociality and those which have several young at a birth, is a phenomenon which can be more easily described than explained. Herbert Spencer developed the poet Schiller's theory of play as a mode of giving vent to an excess of energy which is available in young animals because they are not faced with the energy-using tasks of food-seeking and caring for their own safety. This superfluous energy flows along the most open nervous channels and results in apparently aimless movements of running, leaping, and kicking up the hind legs. This ingenious theory does appear to explain those capering movements of animals and children which occur in good health after any period of confinement. But play is a much more complex activity than random exuberant movement. Karl Groos (1898) put forward the view that play forms a preparation or education for the serious tasks of life, carrying with it no immediate responsibilities. In some ways this is the opposite of Spencer's theory, for it implies that the tendency to play is an hereditary endowment of great evolutionary value. The period of immaturity becomes valuable to the species in that it provides time for the operation of the tendency to play, which develops and perfects instincts and, what is equally important, gives opportunity for modification of behaviour which may be of service to the individual or the society. This theory, though inductive and teleological and not accounting for all examples of play, does link up with the inference we may draw from that type of play in which the mother takes an active and guiding part. Play becomes education, and the longer the period of immaturity allowing opportunity for irresponsible play, the more complete the education. Certainly the animals much given to play and having a long period of infancy show high mental development. McDougall (1931) emphasizes the impulse to rivalry which is a strong factor in play and which is a modification of the combative impulse, 'the impulse of an instinct differentiated from the combative instinct ... to secure practice in the movements of combat'.

At this juncture none of these attempted explanations of play satisfies me. It is doubtful whether random movements referable to superabundant energy should be given the title of play; they are physical pleasurable activity of the order of a motor-mechanism induced by the wellbeing of the organism. At the same time this boisterousness is infectious, as anybody who has had to care for herds of cattle knows. It may serve as a preliminary to indulgence in co-ordinated play. Groos and McDougall pay too much attention to the mock-combative aspect of play and the play educative of capture of prey. In ungulates play may take exactly the opposite form, that of evading capture. I am inclined to be eclectic until a more complete theory forms in my own mind; to observe that play is in evidence when the participants are in a state of bodily health; that the movements in many forms of play are similar to those which may be made in goal-seeking activity in adult life; that the repetition of movements in play tends to perfect patterns of activity; that play may be parentally induced; that play gives opportunity for variations in behaviour and that it is a social phenomenon in animal life. Lastly, uncritically and anthropomorphically, I believe animals play from an innate spirit of fun. It is a function of the organism making for health as well as arising from it. To refer play to purely sexual origins is ingenious but difficult of adequate demonstration. The sexual differences in play are obvious, nevertheless, and erotic play does occur during the rutting season on the part of the hind. Indeed, it is doubtful whether erotic play on the part of the female can be placed in the same category as youthful and family play.

Play which I have observed among red deer has been of the following types:

Young calves up to three months of age	King-o'-the-Castle. Racing. Tig. Mock fights (pattern never completed).
Families, e.g. hind, two-year-old, and yearling	Tig. Tig round a hillock.
Stags	Mock combat?
Hinds in season	Erotic play.

King-o'-the-Castle is a common game with the young of ungulates. A hillock is used as an objective, and each member of a group of deer calves tries to attain and occupy the summit. Rivalry is certainly strong in this type of play, but there seems no hint of mock combat in the actions of running up the hillock and shoving away the holder of the summit. I watched some hinds and calves in the corrie between Beinn Dearg Mhor and Beinn Dearg Bheag on 8 July 1935, for a considerable period, in which time the calves indulged in the four types of play mentioned above. No form of play continued for more than five minutes at a time, and the mock fights were little more than momentary. King-o'-the-Castle would start by one calf mounting the hillock and occasionally rising on its hind legs. This would seem to serve as invitation, for others would look up, leave their mothers, and run towards the hillock. The hillock was worn by the impress of many tiny feet, and it was obvious that this had become a traditional playing-place. When I say 'traditional' I admit that association of the hillock with previous fun may influence their behaviour towards a repetition of the experience when they pass near it again. But I have seen deer calves come from a distance of 50 yards to their chosen hillock to begin playing, as if their play was premeditated.

Racing among calves and lambs can hardly be accepted at its face value of reaching a mark in front of the other fellow. From my own observations I think this form of play more concerned with putting the greatest distance in the shortest time from a point *behind* the runners. Each participant in the race serves to excite his fellow to get farther away. They leave a point more or less together, but there is no definite finishing-mark.

Tig is the game in which one member of the group chases the others till he has touched one of them, when that member does the chasing. At least, those are the strict rules under which children play, but the deer calves play with greater freedom. The chaser may become the chased before he has tigged his fellow with his forefoot. The circle of 10 or 20 yards' radius becomes vivid with the fast movement of 10 dappled elves, and the eye cannot follow each and all of them. Then the game has stopped as suddenly as it began and the players have scattered to their mothers for a drink. I have seen nests of mice acting in a similar manner.

I doubt if mock fights are worthy of that name. The calves stand on their hind legs and make as if to strike each other with their forefeet, but they never do. Two, three, or four may be on their hind legs together, their little ears sticking straight up into the air as if they were really cross.

Tig round a hillock by a family of three or four deer is very interesting to watch. The obstacle provides the opportunity for doubling, a practice which is common in the play of dogs round a house but uncommon among animals lower in the scale. I suggest that this behaviour involves *insight* in the sense used by the Gestalt school of psychologists and by Köhler in his work on the behaviour of apes. McDougall (1931B) has interpreted the word in its psychological sense as meaning 'that the animal does not merely receive and react in reflex fashion to certain sense-stimuli, but that it in some sense grasps relevant or essential relations between the features or objects entering into the total situation'. In the same paper McDougall invites the Gestalt school to go farther and recognize *foresight* in animal behaviour. This is a fence at which contemporary psychology has boggled, for it entails the admission of teleological causation in animal behaviour. Every critical observer is aware that the teleological approach is apt to lead to false conclusions, because the mental state of the animal cannot be grasped, and the end results to which behaviour may lead can be understood only from the human point of view. But that is true of all interpretation of animal behaviour, whatever philosophical standpoint is adopted. We cannot *prove* the conative basis of animal behaviour, neither can the physiological school disprove it. We can, however, use our common sense and apply Lloyd Morgan's canon to the effect that 'In no case may we interpret an action as the outcome of the exercise of a higher psychical faculty, if it can be interpreted as the outcome of the exercise of one which stands lower in the psychological scale.' Mechanistically-minded observers, assuming *a priori* that the mechanistic interpretation is the simplest, have applied a fundamentally, sound caution in such a manner that they raise barriers before their own mental vision. If the physiological school applies its views in equal measure to human behaviour, then its attitude is allowable as a philosophical outlook which is consistent, but once man is placed outside the scheme the mechanists become as ingenious in complicating simple issues on the one hand as are the teleologists on the other.

To return to our family of deer at play round the hillock: I follow McDougall's argument that the acceptance of foresight is a logical outcome of the acceptance of insight in animal behaviour. The mother hind is standing to one side of the hillock, which is about 10 feet high and 10 yards across. The circumference of this obstacle is between 30 and 35 yards. A two-year-old stag is on one side and a yearling hind on the other. The yearling is chasing the staggie. He is running away round the hillock, 10 yards ahead of the yearling. She turns rapidly in her tracks and gallops in the opposite direction to catch the staggie coming towards her. He is so surprised that he jumps into the air, all his legs stiff, wheels round, and is off in the opposite direction. The yearling follows 10 yards before she doubles again and catches him coming towards her. The staggie is not so surprised this time. He wheels instantly and is off. This happens three times more, when the mother hind, who has watched the game attentively, runs to the hillock and joins in to chase the staggie. She runs, perhaps by chance, in the opposite direction from the yearling, and the staggie finds himself between the two of them. He leaps up to the summit of the hillock and the game is over. And then all three romp away 50 yards and begin grazing. The hillock was high enough and wide enough to hide the deer from each other. The young hind's goal was to 'catch' the staggie, we may presume, not merely to chase him. Her turn in the opposite direction as soon as he was out of sight ahead was an advantageous move towards this goal and involved a reversal of the obvious course. The action showed not only insight into the problem but foresight of a mode of solution. This particular incident of play took place in Coire Mor, at the foot of the slopes of Ruigh Mheallan, on 23 May 1934.

Stags play much less frequently than hinds. I have often seen two young stags with their heads low and their antlers together, turning this way and that and occasionally lightly shoving a little. The *tempo* of the whole procedure is slow and the incident may continue for a quarter of an hour. This causes me to wonder whether it is play at all or merely a pleasurable experience. It may take place at any time between September and March when the antlers are hard and present.

Erotic play by the hind which acts as an excitant to the stag is of a different character from those types described and is referable to a different origin. The only examples I have witnessed will be described in the chapter on reproduction.

General Remarks on Sociality; and Comparisons

I have said already that I consider red deer to have attained to a high degree of sociality among the hoofed animals. It is tempting to compare their social system with that of other species not far removed from them. First of all, the red deer is a numerous and colonizing species in Scotland wherever man allows it to spread. By a considerable reduction in physical size it has survived and overcome the changed environmental conditions from prehistoric and even historical times. Our only other truly wild member of the cervine family is the roe deer, a small and shy woodland creature.

In the Scottish Highlands the roe is not a common target for the sportsman's gun, and in many places – everywhere on the ground I have worked over – these little sprites are encouraged. There are probably less than a score of them in the Dundonnell strath and they seem not to vary in numbers. I have watched them with interest whenever opportunity has allowed, and their habits offer a striking contrast to those of the red deer. Roe deer are seen only in twos and threes, and it is extremely difficult to predict whereabouts in the woods they are likely to be. In fact, they are constantly making considerable shifts of ground and observe no strict daily rhythm. If these small families of roes are watched carefully it becomes obvious that there is no matriarchy here. The buck is the leader of the little band for most of the year, though he may go off to the hill by himself for a month or two in high summer. He is often in front as they graze along, and it is he who makes those sharp glances among the trees and sniffs the air in delicate manner. His mate and the fawn or yearling follow, and do not show the nervous anxiety of the red deer hind. The buck is also a very pugnacious fellow and will fight at periods other than that of the rut. The does drive away their young ones when they are yearlings if new calves are born then, and the buck will not tolerate the presence of another male in the group beyond the age of a yearling. We see then that there is a lack of that family cohesion which is so characteristic of the red deer. This smallness of the group and comparatively early expulsion of the young must be to the detriment of the increase of the species in a country where it has to contend with predators. In Scotland foxes and eagles take toll of the fawns, and in Dundonnell I think the wild cat does also. It will be obvious that the underlying point of difference is that the red deer is matriarchal and the roe deer patriarchal, and this is a clearer concept than thinking of one species as being gregarious and the other not. Matriarchy makes for gregariousness and family cohesion. The patriarchal group can never be large, for however attentively the male may care for his group he is never selfless. Sexual jealousy is always ready to impinge on social relations leading to gregariousness. As Hobhouse (1913) has said, 'The principle of force is the very antithesis of the principle of social ethics';

and I contend that the matriarchal system in animal life, being selfless, is a move towards the development of an ethical system. The governing male cannot keep more females than he is physically able and he will not allow other males to mix with his group. Family leadership by the male sets a definite limit to sociality in a species in which the male displays sexual jealousy. Matriarchy, which entails separation of the sexes for the greater part of the year and relieves the male wholly of parental care, is a decided social advance which helps the species towards survival. A clear idea of the complex of reproductive processes and territorial state of the sexes outside the breeding-time is fundamental to an understanding of animal sociality.

The goat is also wild in the area over which I have worked and flocks are interesting to watch. At some time the stocks of wild goats here must have come from domestic animals, and there is one herd of six beautiful white goats in the Glas Thuill which is descended from two or three of the native type which escaped 10 years ago. Those on the Gruinard ground are black, black and white, and grey. The social system of the feral goat appears to be patriarchal, but the leading male tolerates the presence of younger males. There seems also more co-operation between the several individuals of the group. The goats live in the more inaccessible parts of the ground, and though they will not allow the near approach of man their retreat from danger is unhurried and remarkably complete. The farther away a man may be, the more he can see of their movements. The goats do not move. The man comes to within 200 yards. The goats watch him in their own laconic fashion. Then, probably, in a rocky country he must pass out of their sight before he can approach nearer. That is when the goats disappear. Like the deer, they have an objection to allowing an approaching human being out of their sight. When the herd of goats retreats into the rocks the kids go first, their mothers directly behind them and giving them a nudge with their foreheads now and again to guide the kids in the right direction. The males bring up the rear. The whole movement is orderly and very pretty to watch. Goats seem to have conscious knowledge of their tactical position in relation to an approaching human being which influences the extent of the movement of the group.

Here is a story where the laugh is against me. I was in the birch-wood near the foot of Loch na Sheallag. There was a slight sound in the rocks above me, and looking up I saw, a black-bearded face peering over a ledge at me. The goat's amber eyes were calm and dispassionate and his lower jaw moved regularly across and across as he chewed his cud. He being a goat and so nonchalant, I could not resist climbing up the rock to move him. He went on chewing his cud. The nearer I got to him the more directly below him I was, and his interest compelled him to stretch his neck a little farther to watch my movements; but he did not get up or stop chewing. When I was less than 20 feet below the ledge I came to a pitch I could not climb, so there I stood looking up at the black face, the regularly moving jaw, the amber eyes, and the ringed horns. I should exaggerate if I were to say he was grinning.

Hill sheep in Scotland have little social system. They have marked territorial preferences which are embodied in the term 'hefting', and individuals of the flocks have places on the ground which they like particularly. In Scottish hills man has been at pains, unconsciously perhaps, to breed out the social instincts of his sheep. The Mountain Blackface breed feeds wide and does not collect in knots of more than five or six. Man wants them to feed wide in a mountain country where there are no serious predators, and he is their social overlord when he wishes to move them. In Spain the Merino sheep of olden times were known as the 'transhumantes' because they had to make long journeys in large flocks between winter and summer grazings. This flocking instinct, which is of a genetic character, was fostered for ease and safety on the journeys, and to this day sheep of Merino lineage flock together naturally and feed over the country as a flock. This instinct is made use of to-day in the newer pastoral regions of the world, where territories are large and predators numerous. There is, for example, a great deal of hill land in British Columbia too hard for sheep of Merino ancestry and yet perfectly suitable for Blackface sheep. But the presence of wolves and cougars prevents the wide-ranging Blackface being ranched there.

Whatever the flocking instinct of sheep, man has discouraged any expression of leadership among them, for individual social leadership by sheep is most likely to conflict with the wishes of the shepherd.

It has been my good fortune, though I did not look upon it in that light at the time, to work with a flock of brown Shetland sheep in which there was a fair infusion of Soay and Moufflon blood. These sheep were akin to the primitive or wild type and showed a high degree of individuality. They could not be worked by a dog and there was no flocking instinct. Two or three ewes and their lambs would go away on their own, and it was obvious how one ewe of each group would take the lead. I saw the same kind of behaviour among the pure-bred Shetland sheep on the cliffs of the North Isles of Shetland. When moving a flock of sheep the shepherd does not bother much if one breaks away. It soon comes back of its own accord. If a wild moorit breaks away, it keeps on going.

Conclusion

We have noticed that the territories of red deer in West Highland forests are not more than a few miles long. On the Central Highland plateau I believe they are much longer, but from the knowledge we have of wild ungulates in other parts of the world, the territories of our red deer are inconsiderable in length. The caribou of Northern Canada trek in large herds a distance of some hundreds of miles, and the woodland bison has a 400 mile yearly range. Ours is an insular climate and there is not the meteorological necessity for long treks. The long annual trek may reduce a territory to the shape of an attenuated dumb-bell, only the ends being comparable with the territories we have been considering. I have been told that some herds of elephants will make an annual trek in the shape of an oval each year. Such a territory would seem to be of a wholly linear type, the periphery of an immense letter O. The main reason for a seasonal trek by animals is to reach a different set of conditions or to maintain a closely similar set. The deer of the West Highland forests are able to achieve this change of habitat most easily by altitudinal as well as horizontal movement. The glens are not much above sea-level, and a height of 3,500 feet may be reached by a movement of three miles in plan. In the prairie type of country such a range of winter and summer conditions as, say, the Carn na Carnach hind territory provides, would mean a trek of many miles.

Scottish red deer may, as I have shown, carry as many as 200 or more individuals in one territory. Such large numbers are always hind groups, and when the groups are as large as this they are rarely all together. They are a sign of an undisturbed population long settled and ranging over a large territory by tradition. Their sub-groups of 30s and 40s are more comparable with the smaller groups found elsewhere.

When we read of the very much larger herds of caribou and of other species of the cervine family in other parts of the world, such as North America and Siberia, we may wonder why our own deer are found so often in small groups. I think the answer lies in the same reason which makes the sheep of Merino ancestry more suitable for the Canadian ranches. Animals as a large group are more immune from the attacks of carnivorous enemies, and straggling twos and threes are their easy prey. Snow is less of a danger to a large mass of deer than it is to small groups, and snow in our country is the most potent influence on flocking. Adult deer in Scotland, as I have pointed out already, have no carnivorous enemies, and perhaps this freedom during 300 or 400 years has broken the habit of large groups keeping close formation. We read of 'tainchels' in earlier times when many hundreds or thousands of deer were driven into narrow passes or highwalled corries. I think such an event would be most difficult to stage nowadays even with a large number of men. Not that there are less deer: there are possibly more: but I think the small groups we know to-day would break and double like the Soay sheep and refuse to pack and move as one mass.

Sdobnikov (1935) has described the biological structure of the reindeer herds of the Russian tundra. Herds are large and consist of a central group and a peripheral group. He says that crossing over from one group to the other is rare, and individuals stay either in the core or the periphery throughout life. The leader, the oldest female and always of the vanguard, is highly nervous, and the peripheral deer are more nervous and watchful than those of the core. Sick individuals tend to pass to the outside – a preliminary, one may assume, to becoming solitary. Within the herd are family groups of 30 or more head. Here, then, is a social system closely similar to that of the red deer, but the groups are massed as a herd in face of the more adverse environment.

The observance of definite territories by the red deer of the Scottish Highlands calls to mind Sir Arthur Keith's presidential address to the Royal Anthropological Institute in 1916. His general thesis is that the establishment of territorial rights among tribes of human beings sets up barriers to interbreeding and exerts an influence on the evolution of new races. No minute anatomical measurements comparable with those of Keith on the human subject have been made on different herds of Scottish red deer. English park deer have longer faces and apparently less wide muzzles than wild Highland deer, and the antler type is different. The true Highland antler is thin, very rough, tending to spread and to wave, and I think these characters are genetic in origin though they are influenced by environment. But within the Highland forests there are slight differences which might repay investigation by an anatomist interested in evolution. The environment exerts great influence on the growth of antlers; but, irrespective of size and weight, the taxidermists of Inverness who set up most of the good heads from Highland forests are able in a broad way to guess the forest from which the stag has come merely by antler type. The stags of Harris, an island forest, have small antlers of fine shape and spread; those of Strathconan and Ledgowan in Ross-shire are large, heavy, and well-pointed; those of Langwell and Braemore show a strong English type; and many in the Dundonnell area are badly shaped, close set, and of few points. I shall not go farther into the subject of antler type, for it is outside the scope of this book, but it is evident that differences in this obvious character exist from district to district. Millais (1906) remarked that the Dundonnell deer were the smallest in Scotland, and I am inclined to agree with him. The poor alimental conditions have brought this about, no doubt, but through the generations there will have been natural selection of the small size best fitted to the terrain, and the character is genetic as well as environmental.

We have seen that the hind groups are closely bound by the social system to their own territories, but when the rut breaks the stags leave their grounds and come into those of the hinds, and what is important to note is that at this time stags may travel long distances and may breed far away from where they feed or were bred. My own estimate of the proportion of strange stags on my ground in the height of the rut is 10 per cent of the total number. I am, of course, unaware of the distance these strangers have travelled, except when I have seen a particularly good animal which must have come 30 or 40 miles from his home ground. As the figure I have given means a proportion twice as great of the actual breeding stags, it is obvious that there is a fair amount of outbreeding. Partition of the sexes for most of the year and the urge to travel of the rutting stag prevent the appearance of endogamy in a closely territorial species. But the social system is sufficiently close, wholly so on the female side, to give point to Keith's thesis as applied to red deer.

PLATE IA
Sgurr Fiona and Coire Toll an Lochain
Walter Stephen

PLATE IB
Carn na Carnach – the favourite ground of the deer
Walter Stephen

PLATE 2A
A herd of red deer in winter
© *Deer Commission for Scotland*

PLATE 2B
Stag feeding on heather
Neil McIntyre

PLATE 3A
Hind keeping watch on hill
© *Deer Commission for Scotland*

PLATE 3B
Stags boxing in early May
Neil McIntyre

PLATE 4A
June – Hind and calf
Neil McIntyre

PLATE 4B
Summer – Stag in velvet
Neil McIntyre

PLATE 5A
August –
Stag rubbing tree
Neil McIntyre

PLATE 5B
Deer wallow
Laurie Campbell

PLATE 6A
Stag roaring during the autumn rut
Neil McIntyre

PLATE 6B
The rut – Stag thrashing its head in peat bog
Neil McIntyre

PLATE 7A
The rut – Stags sparring
Neil McIntyre

PLATE 7B
The rut – Stag scenting hind
Neil McIntyre

PLATE 8A
Red deer stag and hinds in a quiet moment
Laurie Campbell

PLATE 8B
Autumn stags in Strath Dearn
Laurie Campbell

CHAPTER SEVEN

Movement: The Influence of Weather

MOVEMENT IS ONE OF the chief means by which the higher animals maintain themselves within the fairly wide limits of ecological normality. The movement of red deer over their territories may be influenced by the following types of factors:

1. *Meteorological*: the factors of temperature, humidity, wind, rainfall, snow, and frost are potent instigators to movement, and some of these may become stimuli in the physiological sense.

2. *Biological*: which are largely factors of disturbance by insects, predacious animals, and man.

3. *Physiological*: embracing particularly the functions of nutrition and reproduction. Food-seeking is one of the animal's main occupations and is responsible for wide movement. The physiological processes of reproduction influence movement in both sexes, but most markedly in the male.

4. *Psychological*: which includes many small everyday movements, the urge to colonization, and to some extent play. Psychological causes of movement, other than those linked with the foregoing influences (1, 2, and 3), may be summed up in the word choice. Choice is a factor never to be disregarded in the study of animal behaviour, though it must be admitted that faulty observation can sometimes allow choice wholly to cover a movement referable largely to other influences. But choice is present in some degree in all movements other than purely reflex ones and especially in those involving curiosity and problems of insight and foresight.

This brief analytical scheme, which will be extended in this and the following chapters, is of value mainly from the systematic point of view and does not stand for a causal-analytical approach or an outlook of connexionism from the writer's mind. His standpoint tends to be holistic or organismal, and throughout the writing of this book the spatial necessity of dividing the work into chapters has been irksome, for no one of them can stand alone or does not overlap the rest. Movement, like the rest of behaviour, is an activity of the organism as a unitary whole; and our classification merely assists in studying particular, but not discrete, aspects of challenge and response. I have tried to show the interrelation and interdependence of factors in the diagram, FIG. 4.

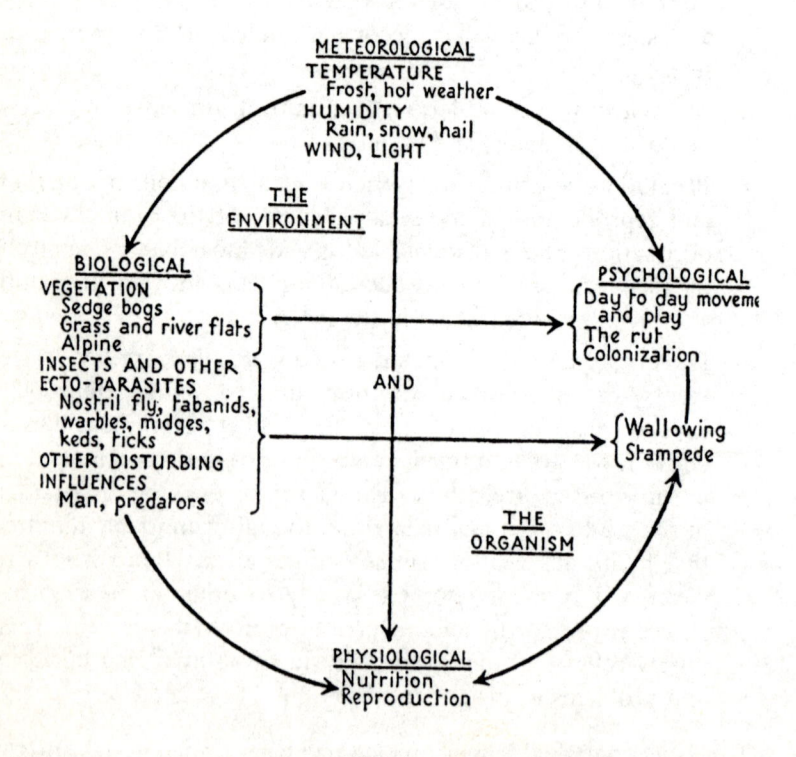

FIG 4
Movement in red deer.

Meteorological Influences on Movement

Temperature. What are the movements of the deer which are made in the course of the year and which are influenced by temperature? We should first of all study the graphs, FIGS. 5 and 6, which show the average monthly maximum and minimum temperatures and the magnitude of the differences between maximum and minimum temperatures month by month. These temperatures, taken at Brae House, are representative for a good number of positions in the forest, but by no means for all. Had the thermometers been placed across the glen on the northern face at the same altitude, the minimum temperatures would have been lower and therefore differences between maxima and minima would have been sharper: in the birch-woods the differences would have been less. At first the thermometers were placed at the standard height of four feet above the ground in an open place but shaded from the sun by a rough wooden screen. It is doubtful whether this is an optimum height for biological work. The height of meteorological instruments above ground must be changed according to the type of animal studied. For deer standing with their heads high, four feet is quite a useful height, but their heads are carried much lower this for the greater part of their time, such as, for example, when they are grazing or lying down. The movements of the thermograph and hygrograph were therefore observed in a screen placed at a height of two feet above the ground in a small rush-thatched shelter built for the purpose. The instruments were sheltered from direct rain and sun, but were otherwise open to the weather. Thermometers were also placed at an altitude of 1,000 feet, north of Brae. It was difficult to duplicate the conditions for setting which would make the figures strictly comparable with those gained at Brae, and for many reasons these thermometers could not be read as regularly as those at the house. Strict comparison was not the aim in view, for the figures would have interpreted conditions in two loci only in a country which offers a wide variety. The figures from the 1,000 feet level provided the information sought, which was to what extent there was seasonal variation in differences between temperatures at 1,000 feet and 150 feet, particularly of the minimum temperatures. From January to the latter end of June the minimum temperature at 1,000 feet is fairly regularly about 5o° F lower than at 150 feet. The lapse-rate of temperature is approximately adiabatic. From July to September the lapse-rate is only 2–3° F, and during this time the minimum temperature at 1,000 feet is frequently higher than at 150 feet, i.e. there is an inversion of the lapse-rate. There is a sharp drop to an average difference of 5o° F again in October. Had the hill thermometers been placed at 2,500 feet I

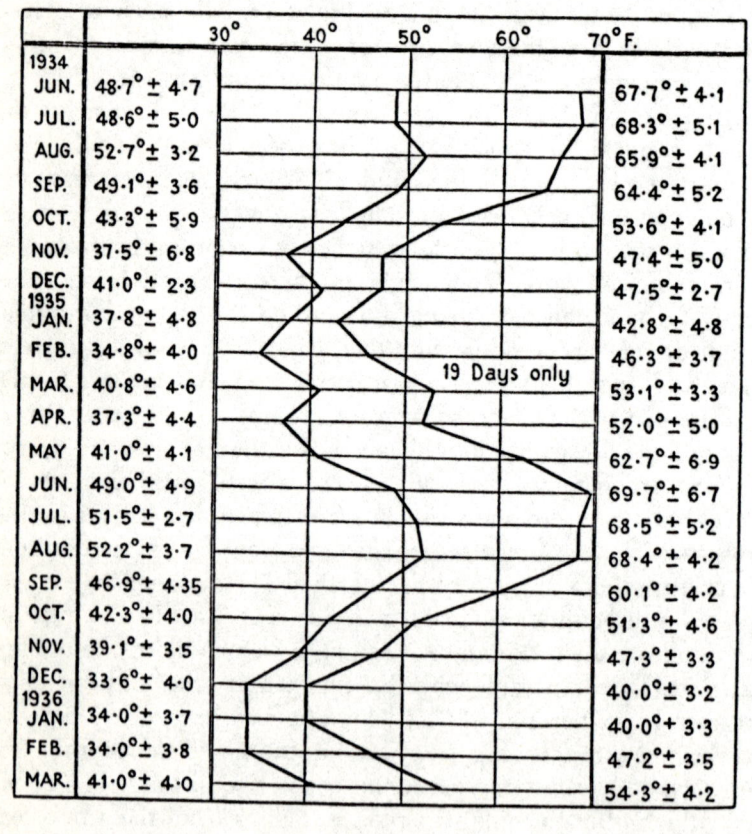

1934	Min	Max
JUN.	48·7° ± 4·7	67·7° ± 4·1
JUL.	48·6° ± 5·0	68·3° ± 5·1
AUG.	52·7° ± 3·2	65·9° ± 4·1
SEP.	49·1° ± 3·6	64·4° ± 5·2
OCT.	43·3° ± 5·9	53·6° ± 4·1
NOV.	37·5° ± 6·8	47·4° ± 5·0
DEC.	41·0° ± 2·3	47·5° ± 2·7
1935 JAN.	37·8° ± 4·8	42·8° ± 4·8
FEB.	34·8° ± 4·0	46·3° ± 3·7
MAR.	40·8° ± 4·6	53·1° ± 3·3 (19 Days only)
APR.	37·3° ± 4·4	52·0° ± 5·0
MAY	41·0° ± 4·1	62·7° ± 6·9
JUN.	49·0° ± 4·9	69·7° ± 6·7
JUL.	51·5° ± 2·7	68·5° ± 5·2
AUG.	52·2° ± 3·7	68·4° ± 4·2
SEP.	46·9° ± 4·35	60·1° ± 4·2
OCT.	42·3° ± 4·0	51·3° ± 4·6
NOV.	39·1° ± 3·5	47·3° ± 3·3
DEC.	33·6° ± 4·0	40·0° ± 3·2
1936 JAN.	34·0° ± 3·7	40·0° + 3·3
FEB.	34·0° ± 3·8	47·2° ± 3·5
MAR.	41·0° ± 4·0	54·3° ± 4·2

FIG 5

Monthly average maximum (shade) and minimum temperatures.
Taken at Brae House, Dundonnell – altitude 150 feet.

believe the results would have illustrated the point better: that during summer there is much less difference between minimum temperatures at altitudes up to 3,000–4,000 feet, that there is a sharp drop in October, and a steady and fairly wide difference during the winter and until June. Sometimes in summer I have camped at over 2,000 feet in order to get clear of the midges. Those nights have generally been warm, and I have wakened early to find myself in sunshine and a sea of mist and colder air below. This inversion ends usually by 10 or 11 o'clock in the morning. The upper slopes

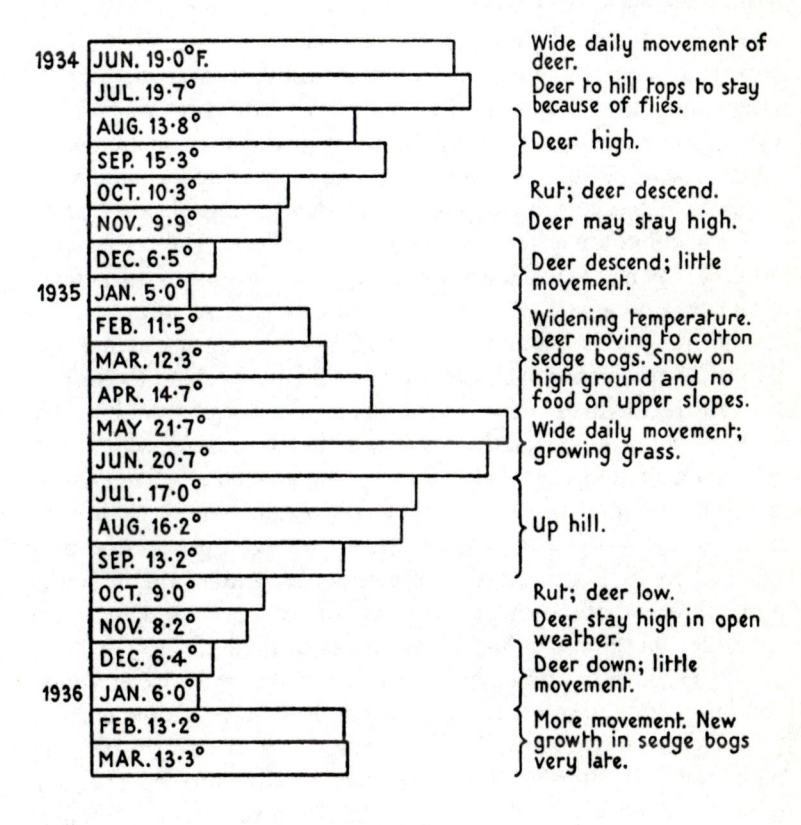

Year	Month & Temp	Notes
1934	JUN. 19·0°F.	Wide daily movement of deer.
	JUL. 19·7°	Deer to hill tops to stay because of flies.
	AUG. 13·8°	} Deer high.
	SEP. 15·3°	
	OCT. 10·3°	Rut; deer descend.
	NOV. 9·9°	Deer may stay high.
	DEC. 6·5°	} Deer descend; little movement.
1935	JAN. 5·0°	
	FEB. 11·5°	} Widening temperature. Deer moving to cotton sedge bogs. Snow on high ground and no food on upper slopes.
	MAR. 12·3°	
	APR. 14·7°	
	MAY 21·7°	} Wide daily movement; growing grass.
	JUN. 20·7°	
	JUL. 17·0°	
	AUG. 16·2°	} Up hill.
	SEP. 13·2°	
	OCT. 9·0°	Rut; deer low.
	NOV. 8·2°	Deer stay high in open weather.
	DEC. 6·4°	Deer down; little movement.
1936	JAN. 6·0°	
	FEB. 13·2°	} More movement. New growth in sedge bogs very late.
	MAR. 13·3°	

FIG 6

Differences between average monthly maximum and minimum temperatures. Taken at Brae House, Dundonnell – altitude 150 feet.

of the hills get the sun earlier in the morning and later at night, and the daily range of temperature is less wide than in the glens. On this score alone there is good reason for the deer remaining high during the months of July, August, and September. The sharp drop, or, technically speaking, the rise in the lapse-rate, which occurs early in October, is evidenced by the sprinkling of snow which usually appears on the hill-tops at that time.

Deer, like many other animals, do not mind cold *qua* cold; they may even seek it; but they are often much affected by sharp alternations of temperature. We shall see that the movement of red deer which is influenced by temperature is always towards the most *even* conditions. This is a very different thing from avoiding cold. Other influences may mask and overcome from time to time this rule of movement in relation to temperature, but in an analysis of behaviour of red deer in Scotland it can be accepted as axiomatic.

A sudden access of cold almost invariably brings the deer down, and it also has the effect of causing them to flock. The families join others in their territorial groups, and, if conditions are sufficiently bad, socially separate groups will join and daily movement is restricted. At this juncture there is no point in stating actual temperatures at which movements of this kind can be expected, for the state of the thermometer cannot be considered apart from humidity and wind prevailing at the time and the conditions which were obtaining immediately before. Daily movement in winter is not generally great; it may be taken to vary normally between a range of a furlong and half a mile. It is least in December and January, when there is least variation between maximum and minimum temperatures and considerable variation between minimum temperatures on the upper and lower slopes of the hills.

Daily range of hoofed animals is never very great except under conditions of migration and emergency. Red deer in summer (after June) do not move much beyond a range of a mile in the day, and often as little as a furlong. It may be of interest to quote from Leopold (1933) the estimated distances of daily movement in American hoofed animals, some of which make long annual migrations:

Elk (Moose?)	⅛ – ¼ mile
Mule deer	⅛ – ¼ mile
Antelope	⅛ – ½ mile
Buffalo	⅛ – ¼ mile
Mountain sheep	1/16 – ¼ mile

We do not know, of course, how far deer move in the dark. From my feeding experience, when I put handfuls of corn down at many places in Glac Cheann, the deer certainly covered a mile or more of walking while picking up the corn, and if circumstances have prevented my putting down the corn one evening I have found their tracks coming over the bridge almost as far up as Brae House. But this movement may have been induced by my corn. Sometimes in snow I have followed tracks for between one and two miles which must have been made in the night, but these were all within a radius of between a quarter and half a mile.

As the differences in daily temperature grow larger, always excepting the operation of other influences, daily movement is over a greater range and it is most striking in parts of May and June, when temperature has, as it were, a free hand and is not masked by the influences exercised by snow, lack of food, and insects. This is the time when the daily maximum and minimum temperatures are at their widest. There may be daily movements of three miles in each direction, from an altitude of 300 or 400 feet to 2,000 or 3,000 and back again in the evening. (This up-and-down movement is evident to a much lesser extent on good dry days in February and March. The temperature in such weather may reach 75° F in the sun on a southern slope during the afternoon, but as soon as the sun sinks below the summits of the hills across the glen, the temperature falls very sharply indeed, and down come the deer from their grazing at the 1,000- or 1,500-foot level to spend the nights at 300 or 400 feet.) The rangy type of movement taking place in June is illustrated by the following timetable of the Glas Thuill stag company's movements during the 10 days 15–24 June 1935.

15 June   In Glac Cheann during the evening (400 feet).
16 June   Most of the day on Carn na Carnach (1,500 feet) and back to Glac Cheann in the evening.
17 June   Across Coir' a' Ghiubhsachain and to the foot of Glas Mheall Mor (1,750 feet). Back to Glac Cheann in the evening, eight o'clock.
18 June   Up to foot of Glas Mheall Mor again (1,750 feet) and down to foot of Glac (150 feet) in the evening, seven o'clock.
19 June   Away to Carn na Carnach, much movement about the Carn during the day. Down on quartzite slabs below Carn na Carnach in the evening, nine o'clock.
20 June   On the slopes of the Glas Thuill corrie (2,000 feet and over). Down to head of Glac Cheann (700 feet) at night.
21 June   Higher up the Glas Thuill slopes (2,500 feet). Down to slabs at night between Glac Cheann and Carn na Carnach.
22 June   Carn na Carnach all day.
23 June   Across Coir' a' Ghiubhsachain into the Glas Thuill. Stayed all night.
24 June   At the head of the Glas Thuill.
25 June   Through the pass at 3,000 feet into Coire Mor.

The same thing is apparent in Strath na Sheallag. The deer will be feeding and resting high on Beinn Dearg or the southern face of An Teallach during the day, and at night the wide bog will be dotted with as many as 200 or 300 deer between Loch na Sheallag and Achneigie. There are other places, too, where deer are sure to be found of an early June evening – the head of Loch Ghiubhsachain, the flats along the Gruinard River, and at the head of Loch a' Mhadaidh Mor in Fisherfield. But in the daytime the deer disappear and take some finding.

There is one physiological reason why temperature should exercise such a marked effect on altitudinal movement at this time.

The deer are shedding their old coats and the new summer coats are much thinner. Moreover, they are wallowing in May and early June, and there must be considerable cooling of the skin by evaporation of water and this cooling may approach discomfort for the deer at night-time. Stags often wallow in the evening and very early morning. Why in the evening? Birds bathe very frequently just before dusk. I can never satisfy myself of the reason why.

The influence of cold on movement in relation to sexual behaviour is profound, and it will be treated in the chapter dealing with reproduction.

Frost may check movement or evoke it, depending on the degree of humidity and ground-conditions. Let us take, for example, the week of frost which followed the wild weather of October 1934 and the heavy snow at the end of the month and at the beginning of November. The weather had been bad enough to imprison the deer in the lower parts of whichever glen they happened to be in at the time, and the main surge of the rutting-season was broken by the torrential rains and sleet-showers from the north-west. Clear skies, still air, and frost are the usual culmination of a stretch of such bad weather. The wind had drifted the snow and left clear patches up to the 2,000 feet level, and the frost hardened the snow sufficiently for it to carry the deer. They went to these clear places to graze each day and could be seen lying on the summits of all the 1,000–1,500 feet knolls in the sunshine. At night they descended to the glens again. Here is a short account of day-to-day events:

1934 23–31 October	Considered the worst weather in living memory round Dundonnell. An average of an inch of rain a day, high winds, and hail-showers. Rut broken up.
1 November	Heavy snow yesterday and to-day.
2–7 November	Deer very low in the straths. As soon as any hint of thaw set in, more snow fell. Snow soft.
7–14 November	Frost came and a still air. Deer became less wary. Deer moved up on the hard snow to places drifted free. Found tracks of stags at

	over 2,000 feet. Missed a lot of stags from Carn na Carnach and the pine-wood. Found them in the Glas Thuill and on Glas Mheall Mor.
16 November	Stags and hinds up to 2,400 feet on Beinn Dearg. Air frosty and very still.
19 November	Thaw set in swiftly with high winds and rain. Deer came down yesterday in a very short time.

This frost and its attendant conditions following a mild if wild autumn invited movement among the deer, and it was obvious that they found pleasure in activity. But what happens more than a year later when, after a bad September and October and an open mild November, sleet-showers come and frost settles solidly on a light snow? There was very little snow on the high hills during November and the deer were grazing almost as high as in summer and staying high at night. Here is a brief account of events:

1935 3–24 November	Open weather, as described, with plenty of sun for the time of year. The deer are moving uphill on the 3–4 November. On 10 November the stags were still with the hinds but were displaying no sexual activity or pugnacity to other stags. Some hinds were getting away on their own, and little groups of spent, older stags were collecting together again on their own grounds. There were slight ground-frosts sometimes at night and the deer were moving about daily at a good walking speed. They were inclined to be playful.
25 November	A most interesting day. The sky was leaden, recorded temperatures did not vary appreciably from those of previous days (maximum 51° F, minimum of previous night 42° F), but there was a very strong south wind of exceptionally even intensity. Relative humidity dropped an average

of 10 per cent from 50–70 to 40–60. The weather felt very cold and I could not stop still for long anywhere. All the deer I saw on this day were timid and irritable, making off at the gallop as soon as they saw me at 600 yards' range, even though I was down-wind from them and in some instances merely walking steadily up the Gleann Chaorachain path, where usually they take little notice at 200 or 300 yards of human beings who keep moving. There were deer in the shelter of Coir' a' Ghiubhsachain and a bunch of half a dozen yeld hinds at 2,300 feet in the most sheltered part of the Toll Lochan corrie. I predicted snow, but no deer were coming down. The barometer dropped nearly half an inch from 29.9 inches.

26 November The snow was down to 1,700 feet this morning and there were hail-showers all day with thunder and lightning. By nightfall the snow-level was 1,400 feet. The deer were coming down but they were irritable. There was minute-to-minute variation between 60 and 100 per cent of relative humidity. The barometer fell slightly. Wind round to east and north.

27 November A rough night of thunder and hail. The snow was down to 900 feet and there were 100 deer on Achachie. Relative humidity was constantly variable between 70 and 100 per cent. Wind north-west.

28 November – 1 December Similar weather and much hail and thunder and lightning. Temperatures between 30–40° F. The wind was from the northern quarters and the barometer was down to 28.5 inches.

2 December Light snow down to the house at 150 feet. The

Carn na Carnach hinds were off their beat and down on the Gleann Chaorachain quartzite slabs. There were two young stags roaring desultorily on Achachie but they were showing no other signs of sexual activity or animosity towards each other. I could find no tracks in the snow above 800 feet. There was a fair number of deer, about 100, in Strath na Sheallag. I expected more. A movement of stags down to the coast of Gruinard Bay was noticeable, involving a trek of from three to five miles.

4 December – 3 January

A succession of dull frosts and occasional light rains which caused some glazed frosts. The deer were not going above 750 feet and their physical movements were slow and listless. They were comparatively indifferent to the human observer in that he could pass by those that knew him, or he could walk on a footpath at 60–100 yards' range without their moving. Strath na Sheallag had filled up with deer and there were between 500 and 600 of them along its whole length. Daily movement was not more than a furlong or a quarter of a mile. The frost having come on a light snow made food hard to come by. The Carn hinds were having to spend much of their time on the Gleann Chaorachain quartzite slabs or in Glac Cheann, but they went back to Achachie whenever there was a slight break for the better in the weather. Snow fell heavily on 23 and 24 December and there was a determined movement to sheltered places which I shall describe later. On the morning of 24 December the snow was beginning to drift badly under the stress of a cold east wind. I attempted unsuccessfully to reach Strath na Sheallag, but reached the summit of the Gleann

Chaorachain track. There I found a two-year-old staggie wandering apparently aimlessly. He was unhappy and alone, and when he saw me he went into a panic. He leapt along across the bog, the surface of which was now smoothing rapidly with the drifting snow. The staggie floundered about in the still soft drifts, but went on and on till he disappeared from sight in the wraiths of driven snow. What was he doing up here all alone? Something had been wrong. He had lost his group and everything was adding to his fright and accentuating his inability to fend for himself. The bog contained many lochans, half covered with ice and snow. I feared for this youngster's safety if his panic persisted. A thaw followed on 28 December and then frost again, but the deer did not move out of the sheltered glens.

4–9 January

Fine open weather. The Carn hinds went back to their own grounds, ranging freely over Carn na Carnach and across Coir' a' Ghiubhsachain. The Glas Thuill stags left the Glac where they had to stay and went back up as far as the lip of the Glas Thuill. The Coire Mhuillin stags which were in the Dundonnell pine-wood along with the hinds went back into their own territory. Round at Gruinard the coast stags could be seen sitting on the tops of the gneiss hills nearest the sea.

10 January –
4 February

An alternation of frosts and snow. The deer were low and did not move far in the glens. They tolerated a fairly close approach from a human being.

It is evident then that under some conditions frost may immobilize herd movements rather than provoke them. The status of frost in relation to movement has to be considered along with that of wind, humidity, and availability of food, and will therefore be mentioned later. Frost also influences oestrus, and therefore movement, in the hind in the same way as it does in the ewe, and this aspect will be discussed in the chapter on reproduction.

Barometric Pressure. Considered apart from its prognostic value when measured by ourselves, I have been unable to notice any influence of barometric pressure on the movements of red deer.

Humidity. The state of wetness or dryness of the atmosphere is one of the most potent influences on movement of deer, and it does not only depress or excite movement in them itself but may affect the organism to such an extent as to intensify the effect of other influences disturbance by man, for example. Broadly speaking, high atmospheric humidity tends to restrict day-to-day movement and a dry atmosphere induces it. But we cannot let the statement rest at that, for a damp atmosphere blankets the effects of the sun, and dry clear conditions allow those sharp differences in temperature between day and night, sunshine and shadow, which we have noted already to have considerable influence on movement. It is impossible to differentiate completely between the two influences purely by observation.

The climate of the West Highlands is a humid one. I have already remarked in Chapter Four that the territorial range of Central Highland deer is greater than that of those in the West, and I have offered reasons for this. The same tendency to long trekking is even more remarkable in the drier atmosphere of continental conditions, and I have mentioned the annual migrations of caribou and other deer in Canada. The alimental conditions provide an overpoweringly good reason for this extended movement, whether in Canada or the Central Highlands; but the weather of 18–21 December 1934 and 8–14 February 1936, and the behaviour of the deer during those periods cause me to think that a dry atmosphere invites movement, irrespective of conditions of food supply and temperature. Both of these short periods were at a time of year when most food is to be found at the lower levels and the general temperature is low. The relative humidity between 18 and 21 December 1934 fell in the

Looking up the Gruinard River towards Beinn Dearg Mhor.
The mass of rock on the left is typical of the Lewisian gneiss formation.

daytime as low as 30–40 per cent at a temperature of 47–50° F, and between 8 and 14 February 1936 it reached the exceptionally low level of 16–40 per cent at temperatures between 35° and 50° F. In the first period the deer were ranging wide to the very summits of the hills and feeding from the club moss on the stones. The hinds were on the move everywhere, even to the extent of leaving their own territories. Donald MacDonald could not find one on the Beinn Dearg beats, there were about 200 in Coire Mor an Teallaich and round about Ruigh Mheallan, and a herd of hinds passed through Coir' a' Ghamhna – an unprecedented event during my observations, and John Cameron did not remember ever seeing them there before. I am not going to say that the dry atmosphere was the chief influence towards this movement, but it intensified it. What is particularly noticeable is that the movement was forest-wide and not confined to one group. I have been told by stalkers that this mid-December movement has been seen before when the weather has been open enough to allow it. Pure observation will not give the solution, and the phenomenon will be mentioned again in the chapter on reproduction.

During the period of low humidity, 8–14 February, the snow was lying thick above 1,750 feet. The deer were ranging the length of their territories below that level and they were inclined to be playful. A period of low humidity in the West Highlands has certainly a tonic effect. I have noticed in myself that the threshold of fatigue is much more distant at such times, and a greater pace on the hills is possible and enjoyable. Our few hens at Brae live in a semi-wild state, and their reactions have been interesting and strictly comparable with those of other animals. When the atmosphere is dry they range as far as 200 yards from the house, fly high into the trees, make a lot of noise, and are generally much more active. On a day of high humidity, even if it is fine weather and still, the hens sit about for long periods with their feathers fluffed out. Squirrels react in the same way. A dry atmosphere sees them out and about the trees round Brae House and quite tame. If it is muggy they may not be seen for a week. I once kept two grey squirrels in a wire enclosure, and the effect of dry atmosphere on increasing their playfulness was striking.

There was another extremely dry week during March 1935, but unfortunately a three-day visit to London, followed by an attack of influenza, prevented my seeing anything of the deer during that time. A slight but steady east wind is characteristic of these periods of low humidity.

Relative humidity has decided effects on the sexual behaviour of the stag, particularly in relation to movement. This role of humidity will be left for discussion in the section on the sexual psychophysiology of the stag in the chapter on reproduction.

Many years ago I asked myself why, given more or less equal conditions of wind and skill, it is much easier to approach deer on some days than others; or, put in other words, why chances could be taken one day which would spoil the stalk on another. I think I might say that from that question grew the desire to do the work of which this book is a record. The question was one of the first to receive an answer, though maybe not the whole one, after the work was begun.

It should be understood that the term 'relative humidity' means the ratio, expressed as a percentage, of the actual amount of moisture in a given volume of air to the total amount of water vapour which this air would contain if it were saturated, under similar conditions of temperature and pressure. Relative humidity, then, in this kind of work cannot be regarded apart from temperature. There is a normal rise in temperature during the day and a fall at night, and the relative humidity is normally in inverse proportion to the temperature. If there is a greater or lesser movement on the track of the hygrograph than would be expected from that of the thermograph and barograph, then there has been a change in the absolute humidity. The conditions between noon on Friday, 7 February 1936, and noon on Sunday, 9 February 1936, are represented in FIG. 7 to show graphically the inverse proportion between temperature and relative humidity. This figure represents dry atmosphere and very slight east wind, conditions under which the deer are relatively approachable. They are calm in demeanour and do not take fright at the slightest hint of danger. It is much more likely that they will remain still and satisfy themselves of the nature of disturbing factors before moving.

Let us take another sample set of conditions, between noon on Friday, 18 October 1935, and 4pm on Sunday, 20 October 1935 – FIG. 8. Here we have minute to minute variation of humidity between 60 per cent, and saturation point until midnight of Saturday, 19 October. Three inches of rain fell during the three day period and there was a high west wind. The

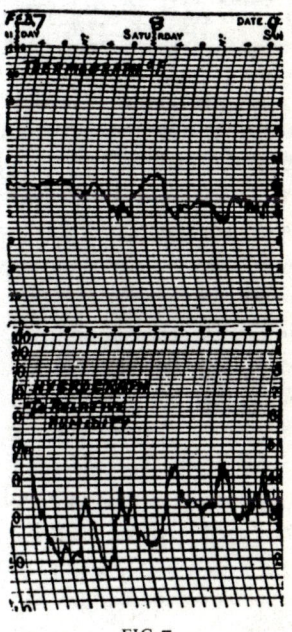

FIG 7

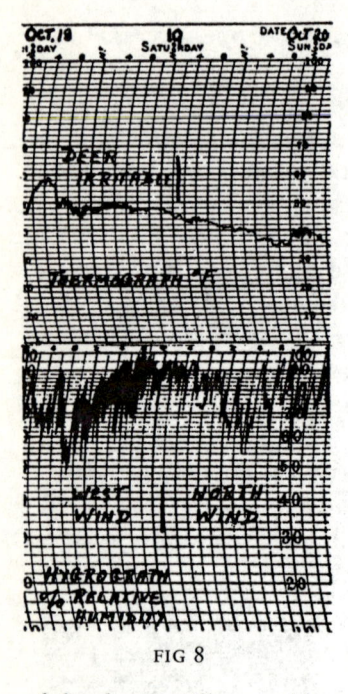

deer were excessively irritable, and the slightest scent of a human being sent them galloping away. The rutting season was in progress, but the deer were in such a state of irritability and discomfort that the harems broke up. After midnight the temperature began to fall, the wind was round to the north, and the oscillations of the hygrograph became much less frequent. By 4 pm on Sunday there was a dusting of snow everywhere, the deer had come down a few hundred feet, they were less irritable, and before the light went on Sunday the 20th I was able to see the stags herding their harems again, and by the following day they were active and the harems discrete. As an example of the marked change in behaviour of the deer I will record how I walked up to within five yards of a stag in Gleann Chaorachain at about 5.30 pm on the Sunday, having exercised no stealth in approach. I stood still. He turned and saw me, interested. He came a step nearer, moved his head sideways on his neck and stood still for half a minute. Keeping his eye on me he walked round me at the same radius to get my wind. Then, of course, he was away. I am sure this would have been impossible two days before, but isolated examples of approach and retreat behaviour are never wholly satisfactory as evidence, for other conditions are never the same in each case. There can be no doubt at all, however, that high and variable humidity makes for constant olfactory stimulation which renders the deer more perturbable. In the course of two years I have had many instances of this type of behaviour to support the contention.

A warm moist atmosphere is a better conductor for scent than a dry one or a cold moist one, and the olfactory apparatus of animals needs moisture in its vicinity if it is to function well. In a moist atmosphere, then, below saturation point, there is greater chance of scent reaching an animal, and scent may reach an animal from a greater distance than in dry air. Also, the same class of scent reaching an animal's nostrils from the same distance under different degrees of humidity will give different strengths of stimulus to the olfactory centres and will result in different degrees of the type of behaviour induced. The conditions of humidity represented by the minute to minute oscillations of the hygrograph are providing constant stimulus to receptivity by the animal, and the effects are particularly noticeable in the periods in between the showers. If there is a stepped increase in humidity there will be a comparable stepped reaction of the deer to, let us say, human approach. If humidity remains steady there is olfactory accommodation and irritability is lessened. Such a set of conditions is represented diagrammatically in FIG. 9.

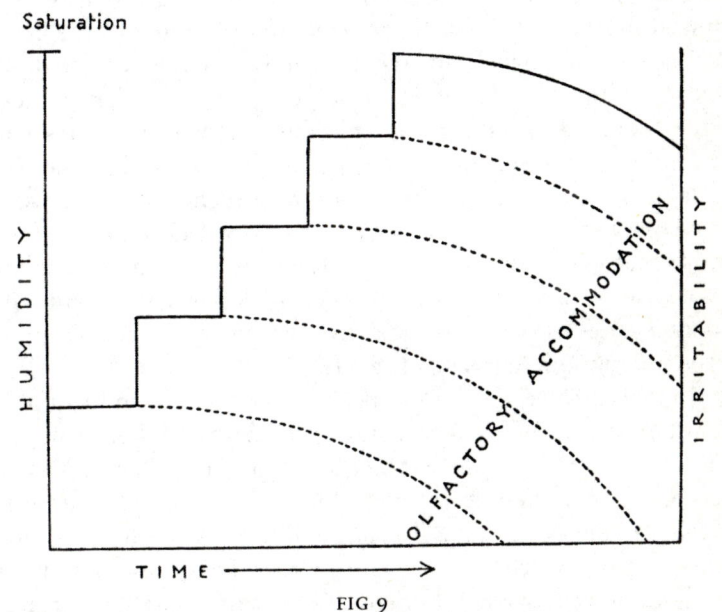

FIG 9

Diagram to show irritability to movement by disturbance in relation to humidity of the atmosphere.

I have said earlier in this chapter that temperature and humidity to some extent cannot be considered apart in their relation to movement. High humidity, we have seen, makes for little movement among the deer, but in the conditions described in the foregoing paragraphs high humidity causes stronger reaction and possibly greater movement in response to the influence of disturbance. The high humidity registered in the evening of each day is not operative in this way normally because the drop in temperature is correspondingly great. Scent does not carry as well on cold air, and in practice it is the general experience to find deer easier to approach after sundown. Here again the factor of light enters into the problem. Also, the air often becomes more still at that time and scent does not travel so easily.

When conditions of saturation are reached deer are more easily approached; they indulge voluntarily in little movement. Saturation may occur under fairly warm conditions, which results in mist, and it can occur during frost when the air is quite clear. Mist offers a physical obstacle to the rapid passage of scent through the air, in that the minute particles of scent oil must become attracted and attached to the particles of water in the atmosphere. Furthermore, mist is frequently attended by conditions of inversion, i.e. the temperature nearer the ground is lower than that above the mist. This results in atmospheric stability, or a low incidence of eddies which would carry scent. There is a third reason why deer should be easily approached and irritability to movement lessened in mist, if my contention is correct that a constantly oscillating state of humidity acts as a series of stimuli to olfactory receptivity. In mist the track of the hygrograph keeps dead steady.

A good example of dry atmosphere, high humidity, and saturation as mist occurring within a very short period was on Wednesday, 13 June 1934, see FIG 10. The weather had been hot and dry, with clear skies and a very slight north wind for several days. Then at 3 pm on this day the sky became overcast and there was a heavy shower giving 0.15 of an inch of rain in half an hour. The ground was warm, the cloud moved away and evaporation was rapid, which brought about rapid cooling of a good layer of air immediately above the ground, and therefore a heavy still mist. This mist continued until the following morning. I had been about the forest during the day and the deer were calm, as they had been for several days. But with the coming of the shower and for an hour afterwards, the few that saw me ran away at a sharp trot from a distance of 400 yards. When the mist formed, thick enough to restrict visibility to within 20 yards, I went on Carn na Carnach and stayed there until after 11 pm Mist undoubtedly helps in approach to deer from the visual point of view in that an observer can walk to within 100 yards over open ground if he makes no sound, but after that it is the senses of the deer other than sight with which he has to contend. There is also the serious drawback to stalking in heavy mist that unless it is known fairly certainly where the deer are likely to be, the observer may stumble upon them, and frightened deer which are unable to smell, see, or hear the location of what they think to be dangerous are likely to go into a panic and stampede. They fear the unknown as we do.

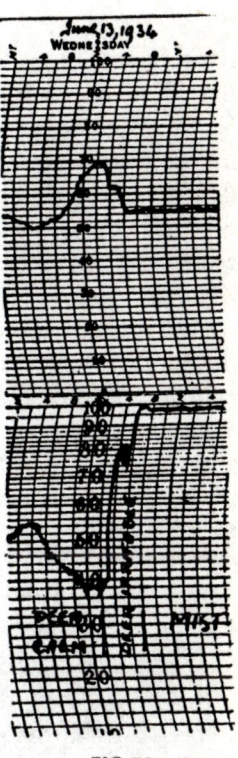

FIG 10

On this night of 13 June I was fairly certain that I would find some of the hinds on the western slopes of Carn na Carnach, and I approached from the summit with care. The stillness of the air was extraordinary. The first sign of my being near to them was the sound of their grazing within a few yards of me, the breaking of the grass as they pulled, the few chews to each mouthful, and the sound of swallowing. I was also within 20 yards of a hind and calf without their knowing, and I got away from the Carn unbeknown. As the deer were there the following day it was evident that they had not taken fright at my track if, indeed, they had crossed it during the night.

Let us consider the conditions of humidity during frost and thaw. Frost in the West Highlands means usually a high relative humidity and a fairly steady track on the hygrograph. Steady frost is often accompanied by a still air, so that scent-carrying eddies are not constantly reaching the deer. Sometimes the track becomes dead steady at round about 95 per cent. These are times when the deer are comparatively tame. A thaw sets the needle of the hygrograph oscillating and the deer become irritable immediately. Klemola (1929) records strikingly comparable behaviour in reindeer in Finland. He says that when the deer are corralled in winter the herdsmen have to exercise great care at the time of a thaw as the animals are irritable then and are inclined to break down the corral and stampede. FIG 11 shows the thermo-hygrographic records of a period of steady frost, a thaw, frost again, and a slight thaw with snow, from noon on Tuesday 17 December to midnight on Sunday, 22 December 1935.

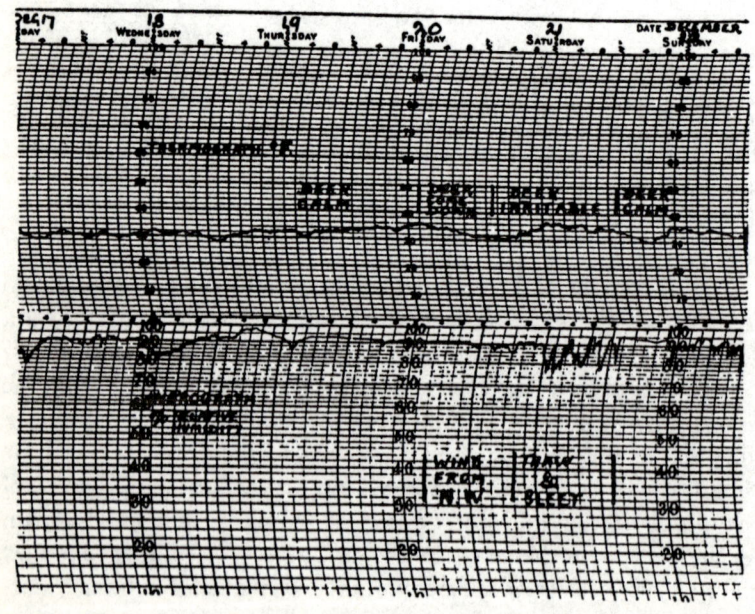

FIG 11

The behaviour of the deer was exactly that which would have been expected on the grounds set forth in this chapter – very little daily movement, deer approachable, deer very irritable (my few feeding deer ran up the steep slope out of Glac Cheann when I took their corn), the snow settled and they quietened.

Wind. Red deer tend to move up wind. It is no more than a tendency, but I believe it to be one which has been much exaggerated by sportsmen and travellers. A west wind of long duration does not overcome the territorial bias of the animals and fill up the West Highland forests with deer. Contrary to expectations, my observations during these two years lead me to the conclusion that wind is not one of the most potent influences on movement. Wind is, of course, the chief factor a stalker has to watch in approaching deer, and its importance from this point of view has perhaps magnified in the human mind the influence of wind on movement of the animals as a whole. Wind can never be considered apart from its quarter, the season, other atmospheric conditions, and the particular piece of country. Deer are always conscious of wind, and it is obvious to anybody who has watched them for any length of time that they depend on it and utilize it as a vehicle of sensory information. We shall have cause to refer to wind again in the chapter on the senses, and only the general impressions are given here.

A south wind of slight or moderate strength will draw deer on to the southern face of a hill. A light north wind in summer will draw them on to the northern slopes. Thus stags may cross from side to side of An Teallach, from Coir' a' Ghamhna and the Loch na Sheallag face into Coire Mor, or from Coire Mor into the Glas Thuill and vice versa, but I repeat that the tendency is not to be exaggerated. A slight north wind in summer provides the best conditions for approaching deer. It usually means a clear sky and a dry atmosphere. The sun is behind the stalker and in the face of the deer, and that is a very great advantage. A north wind in winter invariably brings the deer downhill. West winds are humid and for that reason may cause irritability to possible disturbance and some movement into the wind. An east wind makes for dryness and wide variations in daily temperature and therefore tends to cause greater daily movement. If a wind is strong enough to reach gale force, whatever its quarter and whatever the season, it will bring the deer downhill or into sheltered places.

The vagaries of wind in mountain country are incalculable. There is almost always some degree of turbulence in the corries and steep places, and the small movements of deer day by day, especially in summer, bear a positive correlation with wind conditions in the particular places where they are, in so far as the wind is a vehicle of information and therefore a protection – and not a source of discomfort. There is one type of wind on the high hills in summer worthy of mention, which must be responsible for much movement if a source of disturbance is present. This wind occurs on hot days when the air for a few feet above the ground is dancing. It has no particular direction or quarter, it is exceptionally strong, its path is very discrete, and it occurs only in the ground layer of air for a few inches up to two feet in height. These little winds can be heard racing through the grass near by, causing almost a scream of sound, and if conditions are favourable their erratic course can be watched by the grass or heather bending before them. Perhaps they should not be called winds at all but flying eddies of air, comparable with the 'dust devils' of desert countries. I was lying once on the top of Caiseamheall seven yards away from a dub of water. A flying eddy was coming and I could hear it far away. It struck the water and whirled, and its force may be gauged by the fact that a column of water several inches in diameter rose three feet into the air. It was away again in a fraction of a second. Not the slightest puff of wind made itself felt on me. These flying eddies of hot dry air probably do not carry scent-oil for any long period, but their great pace allows scent to be carried a good distance in a very short time. I have seen two or three stags of a company which was resting peacefully 300 yards away throw up their heads and bolt when struck by one of these eddies that had passed me only a very few seconds before. The eddies, of course, bear no relation to the normal tendency of the wind, and they are therefore one of the hazards of the stalker.

Precipitation. Rain does not appear to exercise much influence on movement. Steady rain tends to restrict it. If the rain is very heavy and accompanied by high temperature, the deer will move uphill to the drier slopes above the peat. The effects of rain as atmospheric humidity have been already discussed.

Hail causes much discomfort to the deer. At the advent of a hail-storm the deer gallop from the open places to the lee of rocks and trees and stay there until the shower goes by. They may drop 500 or 1,000 feet in such a manoeuvre, covering half a mile or more of ground. Hail-storms are often accompanied by thunder and lightning in this part of the country, but I have never seen deer seriously disturbed by these last phenomena. Whether lightning at night-time frightens them more I do not know, for I have never seen deer during a lightning flash at night.

The coming of snow has a very remarkable effect on red deer and their concerted movement before a heavy fall provides what I think is one of the grander spectacles of nature among living things. The day, and sometimes two or three days, *before* the snow comes the deer begin to flock. Twos and threes become eights and 12s. An observer at that time sees no deer quietly grazing. They are moving about all the time, joining up with others, and then in strings they begin their slow purposeful walk into the straths. Territorial boundaries do not exist then, and treks of several miles are common. Strath na Sheallag fills with deer from the whole of its drainage area. The Strath Beinn Dearg hinds follow the Ghiubhsachain River down on to the grassy flats of the Gruinard River, the Carn na Carnach hinds come into the Dundonnell Strath, and the pinewood fills with stags and hinds from the northern face of An Teallach. Companies of stags stream through the Gruinard ground and Fisherfield and gather round the coast of Gruinard Bay. Wherever the watcher may be on the day before the heavy snow begins he sees this activity among the deer. The single files are moving down. Each beast's head seems a little farther outstretched on the neck, and they are curiously indifferent to the presence of man. I confess to a particularly soft feeling for the deer at such a time which is something more than pity. In fact, there is a tinge of reproach to my own kind in it. This emotion is not peculiar to me at this time. I have heard a stalker say, on a day when he had been anxious to shoot a hind, 'Indeed, I could not be putting them one inch farther up the hill to-day.' And he did not go out.

The factors which give deer foreknowledge of deep snow remain a mystery to me. They know before we do, and we can usually tell the day before by the leaden sky and the touch of wind from the north. Or a cold mist may hide the tops and drift about at the 1,000–1,500 foot level, with the north-west wind. This foretells snow too. In this country the movement takes place one or two days before the snow comes, but under continental conditions the concerted movement may take place much earlier. Kropotkin (1904) describes:

> a migration of fallow deer which I witnessed on the Amur, and during which scores of thousands of these intelligent animals came together from an immense territory, flying before the coming deep snow, in order to cross the Amur where it is narrowest ... Like migrations were never seen either before or since, and this one must have been called for by an early and heavy snow-fall in the great Khingan, which compelled the deer to make a desperate attempt at reaching the lowlands in the east of the Dousse Mountains. Indeed, a few days later the Dousse-alin was also buried under snow two or three feet deep. Now, when one imagines the immense territory (almost as big as Great Britain) from which the scattered groups of deer must have gathered for a migration which was undertaken under the pressure of exceptional circumstances, and realizes the difficulties which had to be overcome before all the deer came to the common idea of crossing the Amur further south, where it is narrowest, one cannot but deeply admire the amount of sociability displayed by these intelligent animals.

I quote Kropotkin's account of what he saw almost in full, for it is an eyewitness's account of the type of continental migration which must be vastly more spectacular and awe-inspiring than the treks of a few miles undertaken by our Scottish red deer, and these, as I have said, are impressive.

The common factors occurring before snow which I have been able to record are a drop in relative humidity from about 90 per cent down to about 60 per cent and the wind coming in from the north. The temperature under these conditions usually falls but not invariably, and the north wind may be slight. The drop in humidity is a drop in absolute humidity and accounts for the comparative indifference to human presence which I have observed at such times. I am not convinced that these two common factors are the whole reason for the deer coming down before snow, I merely state them as being common to the day, or two days, before the snow falls. The problem remains to be solved.

Red deer will scratch through a thick layer of snow to the herbage beneath, but they frequently make for areas of long heather at such times. It is unnecessary to scrape then, for the snow falls from the green heads of the heather and the deer put their heads on one side and graze among the twigs under the mantle of snow. Snow does not drift so easily in the long heather as it does on a bare sedge-covered place. The deer will scratch themselves a bed in the snow in long heather, but not where snow may drift. They scrape with their forefeet, clear the work occasionally with their hind ones, step inside, turn round slowly once or twice, drop their knees and their haunches, and then only the long ears are visible above the snow.

The deer go back to their normal winter grounds *before* the thaw sets in visibly. There is an air of activity and busyness in the straths, and group by group the deer trek back, their heads a little higher and the ears farther forward as if in pleasure and anticipation. The temperature may be no higher when the deer start their return journey, but the humidity has risen and become more variable. It should be pointed out here that although the deer always react to a thaw, the movements resulting may be directly opposite. I have recorded earlier in this chapter how fine, frosty, still weather attracted the deer high between 14 and 18 November 1934, which *followed* weather of soft snow. They came *down* in face of the thaw. They would have been immobilized by soft snow had they stayed high then. But when they have been kept *low* by snow and a thaw is imminent their road is back to their home grounds, usually uphill to the 1,000–1,500 foot level. Here is a brief record of a thaw in April 1935:

4–5 April   Deep snow, deer very low and off their own grounds.

6 April   The ground white over, air still and the sky blue and clear. Before 7am. I saw some Carn na Carnach hinds walking up the deer-path crossing the Gleann Chaorachain quartzite slabs. Some more could be seen on the summit of Carn na Carnach, so they must have been on the move before it was light. The Glas Thuill stags were moving over the top of Glac Cheann into Coir' a' Ghiubhsachain. I walked over Carn na Carnach and took up a position in the rocks above and east of Achneigie. Strath na Sheallag was full of deer and easily seen in the snow. If they had meant to stay they would have been resting before midday, but they were all on their feet. They were wandering about as very loose groups until 1pm. Then the groups drifted into their formations for protracted movements and it was a fine sight to see strings of deer going up Gleann an Nid, along Beinn a' Chlaidhemh, across the bog towards the western face of Beinn a' Chlaidhemh, and up the grazing-slopes near Eas Ban at the head of Strath na Sheallag. As I came back towards Carn na Carnach a long string of the Carn deer were coming up from the strath in strict single file and went on towards the foot of Sail Liath at about 1,350 feet. Another string crossed the path at the 1,000-foot level and went on to the flats above Eas Ban. There had been considerable thaw by 5pm but the snow was still lying at 1,000 feet. As I came down Gleann Chaorachain I noticed on a knoll on the eastern side a small liver-coloured stag with a white rump. There were three hinds with him. When I put my glass on him I found he was a Sika, or Japanese, stag. There are some of these deer at Corriemulzie near Achnasheen, 30 miles away

to the south-east through the Fannich mountains. He was in Gleann Chaorachain the following day but I never saw him again.

7 April   The snow had nearly all gone during the night.

We see, then, that snow in its onset and disappearance leads to the most spectacular movements of the deer, and there are secrets remaining for us to learn of the animals' foreknowledge.

Light and Darkness. Broadly speaking, it is safe to say that deer restrict their movement in strong sunlight and that the greater part of their daily movement takes place at night. Deer are on the move in the early morning and there is a general rest-period between 10am and 3pm, during which they are often up feeding for half an hour between 1 and 2pm. As evening falls they begin feeding again, and there is another and shorter rest-period in the late evening. This daily rhythm, as I have set it forth, is a generalization, and variations are frequent. I believe movement at night-time is often considerable. Unfortunately it is impossible to follow accurately except after new snow, in which conditions the season and the snow itself have to be considered as possible influences.

From the latter part of July until May deer like sunlight and take the opportunity of lying in it. The high sun of June, however, drives them into shade, where they lie for most of the day. I have heard stalkers put it that the deer disappear in the daytime during June. The harsh sunlight throws strong lights and shadows, and as the deer are lying in the shade they are most difficult to see. With the dryness of the atmosphere, they are not easily frightened out of their habitual lying-places.

The Glas Thuill stags and some of the Carn deer were interesting to watch in their reactions to sunlight during the first 12 days of June 1934, and for a few days at the end of May and beginning of June 1935. These were periods of hot sun, clear skies, and very slight north or east winds. There are small steep rocky places in the hollow to the north-east of Carn na Carnach, marked 1 on the diagrammatic sketch-map, FIG. 3, and the early morning sun of June shines full into this hollow. By 10am summer time, some of the rocky places would be coming into shade; and at the precise moment when this happened some of the stags could be seen trotting over the ridge from the western slopes, down into the hollow and across to these shady places. Some would stand, others would lie down, and there they would stay until three or four o'clock in the afternoon. Day by day this happened and each beast took up the same position, whether standing or lying, which it occupied the previous day. A photograph taken one day and another 10 days later would have shown no more difference than if there had been only 10 minutes between. Exactly the same places were occupied in the 1935 period and by some of the same beasts. The earth is compressed, and where the small roots of the rowans run across the lying-places they are worn through by the beasts' bodies. The places are used at the time when the deer are shedding their winter coats and when the ticks are thick on them. These beds, then, become littered with old hair, and ticks are never far to seek.

Coir' a' Ghiubhsachain is another place where the deer lie very close in the sunny days of early June, and in such a place of broken rock they are almost impossible to find. I have walked the length of the corrie without seeing a beast or putting them out of their lying-places, and then, watching the place at evening, have seen 30 or 40 deer on the move. Sometimes the bed which a deer has chosen in the shade at 10am is in the sun at 1pm. When this happens, the beast rises and trots across the sunny patch into another bit of shade. The two beds belong to the one animal.

I will record one last observation on the influence of light on movement, at the same time laying no stress on the teleological interpretation of what I have observed. Deer are not easy to see grazing against the background of rock and heather and sedge. In the day time during spring and early summer before they go high, the deer keep on this type of pasture for the most part, but just before dusk they come on to the green places which at this season are becoming a brighter green after the olive hues of winter. If deer are on a green patch in the daytime they are visible to us at very long range, but at dusk we become colour-blind and our range of vision much restricted perforce. The green slope of Achachie, below Carn na Carnach, was a good place for watching this type of movement. On many a day from April into June, as the light grew longer so would the deer come a little later to the green, and the time coincided with the dusk.

NOTE

In the course of the census of red deer and survey of forests I am directing for the Nature Conservancy, I have had more opportunity of observing the behaviour of deer in snow, in forests of different type from Dundonnell, Gruinard, and Letterewe. Where the summits are less spiry, and therefore carrying much greater areas of high grazing, I have noticed that the hinds and calves in best condition will often stay high throughout a period of snow, especially when the weather has settled. The deer then work the areas which have drifted clear. The poor and old hinds not leaders leave the group and occupy the lowest possible ground. These are the ones that would have been better removed during the hind shooting season.

CHAPTER EIGHT

Movement: The Influence of Insects and Food Supply

Biological Influences on Movement: Arthropodan Sources of Disturbance

INSECTS AND SOME other members of the Phylum Arthropoda are responsible for well-defined patterns of behaviour and movement in the red deer. Some of these pests are actively parasitic in that a considerable part of their whole lives, either as larvae or imagines, is spent in or on the hosts. Others are purely blood-suckers and they are in contact with the host only for the short period of the attack.

The arthropods affecting the behaviour of red deer are tabulated overleaf. Of all the species mentioned in Table IV, those flies listed in the family Tabanidae exert most influence on movement of the herds. We shall consider their habits in some detail.

The cleg, *Haematopota pluvialis*, is by far the most common of the Tabanidae in Scotland and it is extremely prevalent in the West Highland districts. The fly is about $7/16$ inch in length and the wings are a marbled smoky grey. The life-history of the cleg has been worked out and described by Cameron (1934). The clegs emerge during the latter half of June and they are active until the latter end of August. There is a closely similar species of cleg, *H. crassicornis*, which is not so numerous as *H. pluvialis*. My observations in Wester Ross lead me to the opinion that *H. pluvialis* emerges earlier than *H. crassicornis* and that in the last week or two of the Tabanid season *H. crassicornis* may be more commonly found than *H. pluvialis*, but total numbers of clegs are never so great at that time.

Tabanid flies are most active in bright sunlight in hot, still weather. The cleg appears to have the greatest range in atmospheric conditions in which it chooses to operate as a blood-sucker, but it is sharply within the periods of hot, still sunshine that the greatest activity is displayed. British Tabanids are as eager to bite human beings as other warm-blooded animals, and for that reason one's personal experiences are of value for comparative purposes. I have registered my first bite from a cleg at 4am and my last at 10pm Greenwich time, in Wester Ross, where there is very little darkness during the latter half of June and the first week of July. There seem to be peak periods of activity during the daytime at about 10am and 4pm. Greenwich time. The cleg is found mostly in the glens and up to 1,500 feet. Above that level there are only a few individuals. I have encountered occasional ones at over 3,000 feet. The season of 1934 was a notably bad one for Tabanids. They reached their zenith of activity on the four days 9, 10, 11, and 12 July, and I made counts of bites at different times of the day. It is worth keeping in mind when making these bite counts that only the females suck blood. The males are vegetarian. At the peak periods bites were too numerous for accuracy, but 30 bites a minute is a conservative estimate. Four bites a minute was the average from 12 noon to 3pm Greenwich time, and before 9am and after 6pm there were only occasional bites. The clegs are inactive during the night. These cleg bites cause little annoyance to the seasoned human male, but the female organism is apt to be more seriously affected by the appearance of painful swellings as sequelae.

Clegs are silent in flight and they are able to settle on the skin very often without the host being aware until the prick is felt. The deer are much bothered by them, and keep off the low ground in consequence. But I have never seen clegs stampede red deer. The clegs attack the deer mostly on the face and legs, in the region of the udder and on the rather bare neck of the milk hinds. The individual flying range of clegs does not appear to be great, but they are so numerous that there is no place free of them below 1,500 feet. On the four days of July mentioned above, they were so thick on the low ground of Gleann Chaorachain and Strath na Sheallag as to cause me to keep my lips well closed to prevent them entering my mouth at the peak periods. Ponies in Strath na Sheallag were wild with the irritation and I saw them galloping by the river bank and

TABLE IV

Arthropodan Parasites and Blood-suckers of the Red Deer

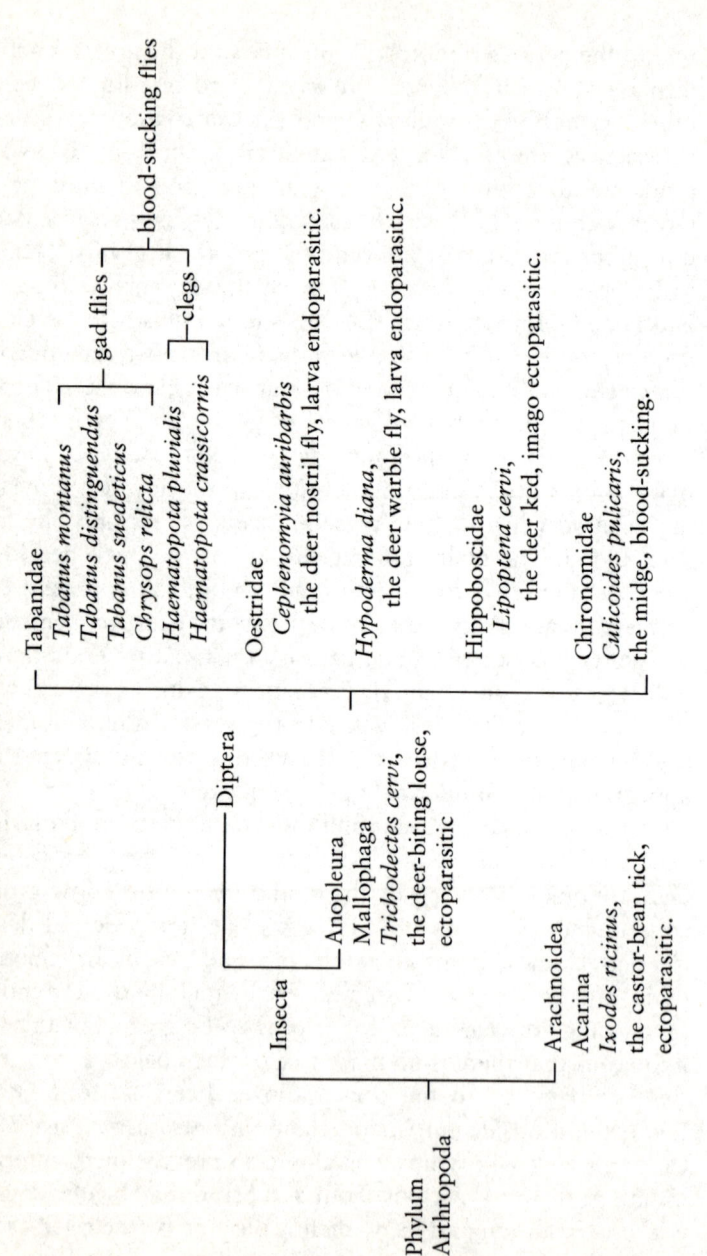

Phylum Arthropoda
- Insecta
  - Diptera
    - Tabanidae
      - *Tabanus montanus* ⎫
      - *Tabanus distinguendus* ⎬ gad flies ⎫
      - *Tabanus suedeticus* ⎭   ⎬ blood-sucking flies
      - *Chrysops relicta*   ⎫   ⎪
      - *Haematopota pluvialis* ⎬ clegs ⎭
      - *Haematopota crassicornis* ⎭
    - Oestridae
      - *Cephenomyia auribarbis*, the deer nostril fly, larva endoparasitic.
      - *Hypoderma diana*, the deer warble fly, larva endoparasitic.
    - Hippoboscidae
      - *Lipoptena cervi*, the deer ked, imago ectoparasitic.
    - Chironomidae
      - *Culicoides pulicaris*, the midge, blood-sucking.
  - Anopleura Mallophaga
    - *Trichodectes cervi*, the deer-biting louse, ectoparasitic
- Arachnoidea
  - Acarina
    - *Ixodes ricinus*, the castor-bean tick, ectoparasitic.

rolling in the water. The temperature was 93° F in the shade and nearly 120° F in the sun. There is a herd of Shetland ponies on the Gruinard ground about Carn nam Buailtean and their behaviour was different from that of those of Strath na Sheallag, probably because they are a bigger group. They massed together and kept up a constant swish of their long tails. The sharp odour of equine sweat surrounded the group and they seemed little bothered. I have never seen our red deer follow this type of behaviour, probably because they go to the high tops where clegs do not go, but Sdobnikov (1935) recounts that the reindeer of the Russian arctic tundra will yard together in large herds as a protection against Tabanid flies. By this massing they tread up and foul an area of ground with faeces and urine, and the odour arising from this and their own body scent acts as a deterrent to the insects. (Incidentally I should describe the body scent of deer, which a sensitive nose can detect down wind or from a lying-place recently occupied, as intermediate between the acrid odour of the horse and the pleasant sweet smell of a cow on pasture. It is a stimulating scent to human nostrils.)

The other British Tabanids are much larger, stronger-flying flies with a greater individual range than the cleg. The mountain gadfly, *Tabanus montanus*, is the most common, but not nearly as common as the cleg. This large fly, about ¾ inch long and with large iridescent green eyes, makes a strong buzz in flight and for that reason can be noticed usually before it rests on the skin for a bite. Its range is from sea-level to 3,500 feet and it is most common at about 1,500 feet. I have never received more than half a dozen bites a day from this fly, and if I had allowed myself to be bitten by them I do not think I should have had more than 20. The bite is more painful than that of the cleg, for it goes deeper and more blood may be withdrawn. Their period of activity in the day is not so long as that of the cleg. They begin at about 8.30am and go to rest by 6pm Greenwich time.

The remarks in the foregoing paragraph apply to the closely similar species, *Tabanus distinguendus*, which can be identified by the purple striping on the iridescent eyes. The largest British Tabanid, *T. suedeticus*, occurs only occasionally. There is another Tabanid,

Chrysops relicta, which is less common than *T. montanus* and *T. distinguendus* and its range is at a higher altitude for the most part than any of the others. Its bite is the most painful of all, and it has the faculty of landing on the skin with very little sound. This is a strong-flying fly with a wide individual range. On 11 July 1934, I felt uncomfortable with a covering of sweat and squashed clegs and went into Loch Gaineamhaidh for a swim. No clegs followed me, but *T. montanus* and *C. relicta* were at my face when I was over a 100 yards out from the shore, so that I was glad to be out of the water again and able to use my hands.

What are the effects of these insect emergences on the behaviour of the deer? The clegs emerged on my ground on 22 June 1934, and by the following day they were numerous and active. We have seen in the previous chapter that there is wide variation in daily maximum and minimum temperatures in June and that the deer make wide movements on an altitudinal scale during the day. The grass on the high hills does not begin to grow until the latter end of May, and in a late year – as was 1934 – there may be very little grazing on the high ground at the end of June. In that year it is probable that the deer would have continued to make their nightly descent to the low ground until well into July, for the high hills were unusually bare. This rhythm was rudely broken by the emergence of the clegs. The deer could be seen moving restlessly on Sail Liath and in the Glas Thuill, but they returned to Carn na Carnach at night, just before dark. They would go back to the hill again soon after dawn.

On 1 July the larger Tabanids emerged and the deer went up the hill to stay. I traversed large areas of the forest in the very early mornings and found no deer below 2,000 feet then, nor any signs that they had been down in the few hours of darkness. The big Tabanids bother the deer much more than the clegs and there can be no doubt that the loud buzz of their approach is an important element in causing the frightened movements of the deer on the high ground. These flies stampede the animals sometimes and are the cause of anomalies in distribution of the deer during the period of the Tabanids' zenith. The stags of the very long Gruinard territory, for example, went into Coire Mor to stay several days before their natural inclination would have prompted them to do so. The grass on the high ground was not well-grown, and it was my opinion that the deer lost condition during the period 1–12 July.

On 13 July there was continual rain. I was on the shoulder of Sail Liath and in Coir' a' Ghiubhsachain during the evening. All the deer were high, but at 7.30pm a trek downwards occurred and was complete by 11pm summer time. It was particularly interesting to watch, in that type classes of deer descended to different levels. Mature stags came down to 1,750 feet and settled on a small grassy alp. A group of yeld hinds stayed at 1,500 feet in Coir' a' Ghiubhsachain below the Toll Lochan corrie. The hinds and calves came right across on to Carn na Carnach at 1,000 feet. A few young stags which were breaking away from the Carn hinds came back on to the Carn also.

14 July was fine and sunny, with high flying clouds and a north-west wind of breeze strength. The deer I had watched the previous night had not moved their positions very much, but they were all lying and grazing *windward* of any possible shelter. On the lee side of Carn na Carnach the clegs were fairly troublesome and there were no deer to be found out of the wind. The deer were still down on 16 July, and the 17th was a day of rain and supersaturation which meant there was no activity among the Tabanids.

On 18 July the deer went high in the early morning and by the following day were settled peacefully in the corries. They stayed there during a fairly active fly period until 24 July, which was a day of lashing rain and high west wind. The Beinn Dearg Mhor stags were down, grazing just above Larachantivore, and round in Glen Muic Beag the deer were on the lower slopes. On the gneiss above Strath Beinn Dearg the hinds were at 1,500–2,000 feet in the lee of any shelter. It should be remembered, of course, that the hinds in that area at no time have more than a little over a 1,000 feet of altitude to work through. On that evening, the 24th, the full complement of the Carn na Carnach deer – nearly 100 of them – were down on the Carn. The milk hinds and calves were down on Achachie, 700 feet, and the yeld hinds a little higher, below the north point. Twenty-five of the Glas Thuill stags were down on the Gleann Chaorachain quartzite slabs at 400 feet.

The weather remained wild, with night temperatures as low as 48° F until 30 July. Needless to say, there was no activity on the part of the Tabanids and it was far too windy for the midges to be out. On 31 July there was a rise in temperature, the weather improved and the deer went high immediately. The deer could be seen moving upwards in the early morning in a purposeful way, not gently grazing their way as they would have done in normal circumstances, but deliberately trekking upwards to get out of the way of the Tabanids and midges which became active during the rest of the day. It is worthy of note that in this type of movement the deer do not wait on the flies becoming active. They move up before the Tabanid flies begin to trouble them. But the midges are particularly active just after dawn. At this time, and for two or three days, there occurred that inversion of temperature between the glens and the high ground which I mentioned in the last chapter, so that there was a dual urge to go to the high ground to stay.

The clegs emerged on 19 June 1935, and the larger Tabanids appeared to become active on 25 June. On 22 June the weather was wet with a strong south wind and a drop in temperature. The Tabanids were inactive and the deer came low. The Glac Cheann hinds came and ate corn I put down for them, which they had not done for over a week. This, however, was the last time in the summer that they came for food. 25 June was a hot Tabanid day and I witnessed a stampede of milk hinds and calves from Carn na Carnach at noon, Greenwich time. This group had stayed on the good green grass as late as possible, but the new season's buzz of *T. montanus* on this sunny day put them away to the hill in complete disorder. The hinds and calves were on the western slope of Cam na Carnach, standing close under the rocks in a very slight west wind. I was just above them watching them shaking their ears and stamping their feet to keep away the clegs. Then came the buzz of *T. montanus*. I was so near to them that they and I heard the same buzz. The hinds threw up their heads and their ears came close together and straight up as they always do when the hinds are annoyed. Some of them turned sharply about this way and that, and then a tiny dappled calf set off southwestwards at a stretch gallop. I think he must have been bitten. It was amazing to see this small fellow, certainly not more than a fortnight old, going at such a pace over the terrible quartzite ground round the head of Coir' a' Ghiubhsachain. He was 100 yards in front of any of the others and never faltered once or changed his step. When hinds move away from disturbance in good order, the shape of the group in plan is like that of a shuttle with the calves close in to their mothers within the group, but in this stampede the shape of the disordered, frightened movement was that of a half-open fan, with the deer moving to the periphery. As I have said, a tiny calf was away in front of all of them. All the hinds were going as fast as they could over the bad ground, probably at about 20 to 25 miles an hour. They kept going for three minutes or more and had covered over a mile. The outside members of the group were at the farthest about a mile from each other. It was possible to see through the telescope that the deer were badly blown. Then began the gathering together again of the group on the shoulders of Sail Liath. This was interesting to watch. The hinds wandered about with their ears far forward, looking here and there and raising their muzzles into the air. The calves, scattered far and wide, did not move more than a few yards but were bleating constantly. As each hind found her calf and twos and threes gathered together, they moved towards Sail Liath. The whole operation of reassembling took half an hour, but the deer were in a distressed state for half an hour longer. They did not graze but stood listlessly, looking sharply about, and the hinds moved their hind legs forward if any calf attempted to take a suck of milk. They then lay down to rest.

By 26 June every beast had gone to the high ground and, as in the preceding year, I made long traverses of the forest late at night and as soon as it was bright in the mornings. There was none to be found below 1,750 feet. In the heat of the day I could see the Beinn Dearg stags walking to and fro at 2,500 feet, nervous and unsettled by the flies.

1 July brought rain and wind and the deer came down to the

1,000 foot level, there to stay for five days with very little movement. They moved up on the early morning of 6 July; the weather was hot and sunny and the Tabanids became active. From the 12th to the 15th the Tabanids were at their zenith in the sultry weather. The air at 3,000 feet was pleasantly warm and light breezes played; the deer were content. The deer remained high through variable weather until 27 July, when a high and gusty west wind came with hard showers of rain. All the deer came lower, but hinds and calves were down to Achachie below Carn na Carnach and in the Dundonnell pine wood. During the morning of the 28th, 9–11 o'clock summer time, the deer trekked up again in purposeful fashion and remained high until 1 and 2 September.

We find, then, how marked is the effect of Tabanid flies on cervine movement. No other species exerts the same herd-wide pressure and produces such spectacular movements of the deer.

Oestrid flies influence the behaviour of the deer in different ways. One of the most unpleasant cervine parasites is the deer nostril fly, *Cephenomyia auribarbis* Mg. This dipteran was first recorded in Scotland by Grimshaw in 1894 and it has been described at length by Cameron (1932a). The fly is larviparous and according to Cameron (op. cit.) each female deposits 500–600 larvae. It hovers round the head of the host and on coming near the nostril ejects a drop of fluid containing one or more larvae. These first-stage larvae attach themselves to the mucous membranes of the nose by strongly reflexed mouth hooks. These small larvae, 1–3 mm in length, remain in the nasal cavities from late May, June, or July until February when, on reaching the second instar, they migrate to the pharynx and increase markedly in length up to 15–20 mm. They quickly reach the third instar and grow to 40 mm in length. In May the affected deer ejects the larvae by violent sneezing. The pupation period is short, 20–30 days, and the flies are on the wing in late May and June. Brauer (cit. Cameron), who described the life-history of this fly in 1863, mentions that the deer do not submit quietly to attack, and yet, for some minutes at least after the larvae are ejected into the nostril it is unlikely that severe pain is felt by the host owing to the tiny larvae becoming attached. The reactions of deer to the approach of this fly are, however, striking and specific in character. *C. auribarbis* is probably more common in some parts of the Highlands than others; in my ground there seem to be comparatively few of them. I have seen them on very few occasions, usually settling on a warm rock, and only twice in two years of watching have I seen deer reacting to their approach. During the winter of 1934–5 I examined several hinds' heads but found no larvae of *C. auribarbis*.

On 31 May 1934, I was on Caiseamheall and saw a two-year-old hind attacked by the nostril fly. I was only 25 yards away from her when I came round the corner of a rock, and although I hid myself immediately it was obvious that she was wholly absorbed with dealing with the fly and I was able to watch at very close quarters. The fly could be seen with the help of the binoculars, hovering and stooping towards the nostril. The hind held her head low and leapt up and down and round and about one place, her legs more or less stiff as she jumped. All the time she was snorting violently and shaking her head. This went on for three minutes, by which time the hind was getting blown. She stood a moment, her head outstretched and low; then she trotted 25 yards away to a dub of water in which she rolled, got up, shook herself and walked 200 yards to join some more hinds. It is significant that when at the water she thrust her nose through it three or four times before lying down to roll. The weather was warm and sunny.

The second occasion I saw the nostril fly and the deer together was on 5 June 1935, in Glac Cheann. The Glas Thuill stags were low in the Glac and on the quartzite slabs following rain, snow on the tops and cold east wind on 3 and 4 June. Seven stags were lying in a group at about noon and they had been there since nine o'clock. One shook his head, rose hurriedly, put his head down, pawed his nose with his forefoot, jumped about like a bucking pony, and stopped suddenly. One of the others began doing the same thing, and six out of the seven were affected in turn. They were now all thoroughly disturbed and they galloped to a dub of water a 100 yards away on the slabs, where they rolled and wallowed. They had settled quietly to grazing again in a few minutes' time, and within a quarter of an hour were lying down to chew the cud. This fly does not cause wide herd movements as Tabanids do, but excites strictly individual behaviour by the animal being attacked.

The other bot-fly affecting deer is the warble-fly, *Hypoderma diana* L., which is on the wing in late May and in June. The egg-laying activity of this fly on the hair of the host does not appear to cause fright among the deer – at least I have never seen them gadding from this source. This finding is in line with that of Cameron (1932*b*). One of the ox warble-flies, *Hypoderma bovis*, causes serious stampeding among cattle, but it is noisy in approach. *H. diana* is quiet and does not alarm the deer. Sdobnikov (1935) says the warble *Oedemagena* has a marked upsetting effect on tundra reindeer herds. The larvae of the warblefly, *H. diana*, presumably burrow into the skin of the deer shortly after hatching and they reach the back in half-grown condition in January. They emerge from the host in April for pupation in the soil. I have never killed a hoodie-crow at this time to examine its crop, but I believe I can say this in the bird's favour, that it eats a good many newly emerged warble larvae. I have seen the hoodies, at this time of year particularly, flying from the birch woods east of Gleann Chaorachain to the deer grazing on Carn na Carnach and Achachie. There they have walked about among them with much chattering and evident enjoyment. Occasionally they would gobble something, and I believe the juicy, newly emerged warbles may be the attraction.

The emergence of warbles must cause great irritation to the deer, and I believe the event to be one of the causes of wallowing in spring.

The deer ked, *Lipoptena cervi*, is a parasite interesting in itself. Unlike the sheep ked, *Melophagus ovinus*, the deer ked is winged in the early imaginal state and there appears to be one generation only each year, whereas the sheep ked may produce several. My observations on its life-history are as follows: it appears as a winged imago in September, a reddish, flat, chitinous creature which it is difficult to push off the skin. Its flight is weak and it seems to tumble on to a host rather than alight. Autumn wallowing on the part of the stags has started by the 10 or 12 September, and the winged deer keds are most commonly found near the wallows and the peat hags where the hinds rub. By the middle of October winged keds have disappeared, though I found one in 1934 as late as 5 December. It is evident that the keds spend but little time on the wing before mating and finding a host. The deer do not appear to react strongly to the onset of these insects. As soon as the keds are in the deer's coat they lose their wings and begin blood-sucking. Their abdomen increases in size and their whole body darkens in colour. They remain on the deer throughout the winter. On stags shot in October and hinds in December and January, and on occasional beasts found newly dead in March, I have found hundreds and even thousands of these parasites. Their debilitating effect on the host must be great. During the winter I have never found puparia of the ked among the hair, but they are present after the end of March. Now the deer wallow at the end of April and during May, when they begin to shed their winter coats. Much of the hair is shed at the wallows and rubbing places, and it would appear that the puparia of the deer ked lie on the ground near such places until they emerge in September, and this would account for their being particularly numerous at the wallows. If my observations are correct, the diapause which seems to occur in the life-history of this species alone of the British Hippoboscidae presents an interesting problem for solution by the entomologist.

The deer-biting louse, *Trichodectes cervi* L, according to Cameron (1932b), is found most abundantly on the deer during the winter months, its numbers gradually decreasing with the advance of summer. I went to particular trouble to find some individuals of this species because I was asked to do so by entomological workers. I examined the skins of a good many deer between October and January but was wholly unsuccessful. Skins were searched with a lens, I shaved patches of skin and washed the shaved hair, but I found no lice on the skin or in the washings. It cannot be assumed, therefore, that the deer in this area are seriously affected by the biting louse, and the parasite has not influenced their behaviour during the course of my observations.

There remains to be mentioned the castor-bean tick, *Ixodes ricinus* L. This parasite is ubiquitous in Wester Ross and will attack any warm-blooded creature. Each year there are cases of red water

above: Stag wallow in Glen Muic
below: Hind wallow in Carn na Carnach

fever among cows in this area, but I have never seen symptoms of the disease among the deer. The ticks come on to the animals in April and May, and they take up positions towards the fore end of the beast, particularly on the throat, chest, and inside the fore legs. They gradually become engorged with blood and drop off the host to the ground. The spring invasion of the ticks doubtless provides another reason for the spring wallowing of deer. Deer calves are born in June, and the few dead ones which I have examined have had no ticks on them. The calves are born after the main invasion and their coats are short and sleek. Lambs are less fortunate in this area. They are born at the end of April and during May, just in the full surge of the tick invasion. Their woolly coats are ideal for the ticks to attach themselves to the skin. In bad tick years, such as 1935, ticks are responsible for an appreciable mortality and unthriftiness among the lamb crop; but, as I have said, the deer calves escape this depressing factor.

In studying the parasites so far, we have found that the host is being subjected to particular discomfort and debilitation in April and May. This is the period of greatest mortality among the deer and the period when both sexes wallow. The warble larvae, fully fed, are coming through the skin of the back, and, what is worse, some are not. These dead ones decay in their cysts inside the muscles of their hosts and must throw an extra strain on the excretory functions of the circulation. The nostril fly larvae have reached their fullest development and are being sneezed out by the deer. The ticks are coming on to their hosts to suck blood, and the keds after feeding all winter are presumably just finishing. Each parasitic species by itself may not cause very serious debilitation, but several species active at the time of year when food has been shortest are responsible for the spring decimation. It would seem certain that springtime wallowing by the deer is wholly referable to the activity of the parasites irritating the skin from within or without. Wallowing assists materially in removing old hair, and it is probable that the peat has of itself a tonic effect on the skin. Peat has long been used in the preparation of soaps and ointments for affections of the human skin.

This attribution to parasitism of the action of wallowing and the movements of the deer which it entails does not account for the difference in wallow type used by each sex, a fact to which I referred in Chapter Three.

We are left with the minute midge, *Culicoides pulicaris*, and its effect on cervine movement still to be considered. This insect emerges in comparatively small numbers at the end of May, and whilst they will suck blood at that time, they cause little discomfort and have no effect on movement then. During the dry sunny weather of June they are not much in evidence, and if there is any wind they lie low in the grass. They emerge in countless numbers in July, August, and September, and the humid, warm, windless mornings and evenings of those months are made unpleasant for warm-blooded animals which are out of doors at an altitude below 1,750 feet. The midges suck blood from man wherever his skin shows; but on deer they concentrate on the eyelids, inside the ears, round the udder, the anus, and vulva, and cause irritation which seems little less to the deer than it is to ourselves. Roe deer are affected much worse than red deer because they live in woods and at a lower level. I have seen them in the evening bounding round and round in circles and burying their heads in bushes or bracken to rid themselves of the clouds of midges. Stags in the Highlands never descend on a midgy evening or morning during the time their antlers are in velvet, for the velvet is a rich and easily available source of blood to the insects, and the tenderness of the growing antlers precludes any vigorous scratching or rubbing. Sdobnikov (1935), writing of the reindeer of the tundra which are unable to get away from the midges as can our Scottish deer, says that the growing antlers may become misshapen in a bad midge season.

Helminth Parasites

It is not the purpose of this book to enter into details of factors in the life of the red deer which do not affect their behaviour. As far as I have been able to observe, helminth parasites do not enter into the consciousness of red deer and do not influence behaviour. T. W. M.

Cameron (1932) and Cameron and Parnell (1933) have treated fully the subject of helminth parasites of Scottish red deer from their own investigations. T. W. M. Cameron's paper discusses prophylactic measures which might be used in deer forests.

Vegetation

The vegetative complex in the North-West Highlands is interesting and especially on the ground covered by the Torridonian Sandstone. Within a comparatively small area, owing to the sharp ascent of the hills from sea-level and the freedom of the upper slopes from peat, there are areas of well-defined associations of plants providing the deer with seasonal types of vegetation. These associations and the effect of climate on the herbage exert an influence on the movement of the deer through the medium of choice or necessity.

I have alluded briefly to the types of vegetation found on particular areas of my ground in the first chapter. The associations of plants are typical of West Highland conditions, but it is worthy of emphasis that the herbage floor of West Highland country is different from the Central and Eastern Highlands. This fact is often insufficiently realized. Heather is not a strong growth in the West, and the main complex is of sedges, bents, mosses, and lichens. The lists which I give below are not intended to represent the whole floor of the area, but those marked associations which concern the lives of the deer.

Strath Beag, Dundonnell, is unusually well wooded. There is the pine wood and a good area of birch-alder-hazel-willow scrub. Pines and larch are used for rubbing, and willow, hazel, and birch are browsed sometimes when food is very scarce during snow, or in May and early June when the young leaves are breaking. All the trees provide shelter. Very few young birches are growing anywhere because the seedlings are grazed each year by deer, sheep, and rabbits. The loss of this beautiful and useful tree is being felt in many parts of the West Highlands and will be felt increasingly in the future. It is incumbent upon owners of hill ground to realize their responsibility.

If small areas of a few acres here and there were fenced against rabbits and deer, the young birches would grow naturally, and in 10 years the fences could be removed. Young alders are affected to a lesser extent, but the number of young trees does not appear to be replacing that of the old ones dying. The old alder wood in Strath na Sheallag is the only wood in the strath and it is much liked by the deer in the worst of the weather. The herbage beneath the birch-alder scrub is modified from that of the bare hill. Bracken or heather grows where the soil is dry enough and the trees not too thick. Bents and sphagnum and other mosses grow beneath the trees in the wetter places, and the sedges appear when the trees are thinly spread. The straths hold a good number of flowers, including kidney vetch, lady's finger, white clover, milkwort, primrose, and tormentil.

Sedges and bents, interspersed with mosses and lichens and with poor heather, form the main herbage of the bare hill below the 1,750-foot level. Above that, on peat free areas, a more truly alpine complex appears consisting of alpine sedges, sheep's fescue, and woolly fringe moss. There are small, flat, boggy areas throughout the forest below 1,500 feet where there is an abundant growth of cotton sedge and yellow carex. These plants provide a green bite early in the year from the end of February to May, and the deer make definite journeys to these flats. The blaeberry is uncommon and grows only in dry situations. The deer are avid for it and the plant reaches fruition only in such places as the deer cannot reach. The crowberry is similarly uncommon and confined to the drier situations on the hillsides.

The movements of the deer in relation to herbage at the time of the autumn snow are interesting. Their movements during November 1934 have been referred to in the last chapter (pp 103–104) and the details will not be repeated. The snow was heavy, there was a certain amount of thaw in the glens, and heavy frost followed for some days. As soon as the thaw occurred on the higher ground the deer went up on to that which was clear, to stay for a considerable period. It was noticeable that without a covering of snow the herbage of the glens had become frosted and withered, but that which had been under the snow at the higher levels during the frost had been protected and was still fresh and succulent. It was therefore far more attractive to the deer, and the herbage, linked with the open climatic conditions, was sufficient to induce them to remain high night and day.

It is noticeable that the deer are eating much *Lycopodium* moss during the autumn, and a rutting stag may be found to have little else but mosses in his paunch at this time of year.

Plant Associations and Movement of Deer

Straths and below 500 feet

Trees: rubbing, shelter, browsing.
  Pine, *Pinus sylvestris*. Strath Beag, Gruinard, and occasional trees.
  Willow, *Salix spp*. Occasional trees and bushes.
  Birch, *Betula alba*. Strath Beag, Loch na Sheallag, Gruinard River, and occasional trees.
  Alder, *Alnus glutinosa*. Strath Beag, Strath na Sheallag.
  Poplar, *Populus tremula*. Occasional small trees.

The Herbage Floor. Food.
  Bent grasses and Fiorin, *Agrostis spp*. General distribution.
  Sweet vernal, *Anthoxanthum odoratum*. General distribution – an early grass.
  Meadow grass, *Poa pratensis* and *Poa annua*. On better and drier ground.
  Crested dog's-tail, *Cynosurus cristatus*. On better and drier ground.
  Purple Molinia, *Molinia coerulea*. Damp places.
  Sedges, *Carex spp*. Not so general in straths as on slopes.
  Woodrush, *Luzula spp*. General distribution and on slopes.
  Tormentil, *Potentilla tormentilla*. General distribution through herbage floor.
  Bedstraw, *Galium saxatile*. General distribution.
  Tufted vetch, *Vicia cracca*. Occasional – liked by deer.

Meadow pea, *Lathyrus pratensis*. Common – much liked by deer.
Bird's eye, *Euphrasia officinalis*. Common.
Milkwort, *Polygala vulgaris*. Generally distributed through herbage.
Yarrow, *Achillea millefolium*. General – liked by deer.
Bell heather, *Erica cinerea*. Patches in dry places.
Cross-leaved heather, *Erica tetralix*. Common, but not general.
Heather, or ling, *Calluna vulgaris*. Patches here and there.
Mosses, *Sphagnum, Polytrichum, Aulacomnium spp*.

Plants occasionally eaten by deer.
Common rush, *Juncus communis*. Patches, river flats.
Furze or whin, *Ulex nanus*. Drier places.
Bramble, *Rubus fruticosus*. Edges of woods, not common.
Nettle, *Urtica dioica*. Where human habitations have been; eaten by deer when dead.
Dandelion, *Taraxacum dens-leonis*. Not common.
Coltsfoot, *Tussilago farfara*. Gravels near rivers.
Ragwort, *Senecio jacobaea*. Patches, river flats.
Marsh and Dog Violet, *Viola spp*. Sheltered situations, woods.
Wood sorrel, *Oxalis acetosella*.

Plants more obviously present but not normally eaten by deer.
Primrose, *Primula veris*. Sheltered situations, woods.
Scabious, *Scabiosa arvensis*. Not common but general.
Silver weed, *Potentilla anserina*. Common in hard, dry places.
Goose grass, *Galium aparine*. Amongst bracken in woods.
Teasel, *Dipsacus sylvestris*. Occasional.
Marsh thistle, *Carduus paulustris*. Common in woods.
Spear thistle, *Carduus lanceolatus*. General but not common – straths only.
Orchis, *Orchis spp*. General.
Bracken, *Pteris aquilina*. Patches in dry places. Large areas on northern side of Strath Beag, Dundonnell.
Bog myrtle, *Myrica gale*. Common all over low ground and on wet places to 500 feet.

Buttercup, *Ranunculus spp.*
Celandine, *Ranunculus ficaria.*
Wood anemone, *Anemone nemorosa.*

Hillsides, 250–1,750 feet

The herbage floor comprising the main food supply of the deer.
Sweet vernal, *Anthoxanthum odoratum.* Commoner lower reaches.
Fiorin, and other bents, *Agrostis alba, Agrostis caninum.* Abundant.
Wavy hair-grass, *Aira flexuosa.* Common, goes high.
Early hair-grass, *Aira praecox*
Silvery hair-grass, *Aira caryophyllea*
Sedges, *Carex vesicaria?*
    Carex ampullacea
    Carex limosa
    Carex glauca
    Carex distans?
    Carex flava
    Carex pallescens
    Carex filiformis
    Carex pilulifera?
    Carex praecox
    Carex caespitosa
    Carex saxatilis?
    Carex pauciflora
    Carex dioica
Deer's hair, *Scirpus caespitosus pauciflorus*
Woodrush, *Luzula spicata*
Rush, *Juncus squarrosus*

} GENERAL DISTRIBUTION

Tormentil, *Potentilla tormentilla.* General below 1,000–1,500 feet.
Milkwort, *Polygala vulgaris.* General below 1,000–1,500 feet.

Heather, *Erica spp*. Patchy as main plant of an area, but thinly distributed generally. (Includes what is called 'bell heather' in Scotland.)

Ling, *Calluna*. Patchy as main plant of an area, but thinly distributed generally. (The 'heather' of Scotland.)

Cotton sedge, *Eriophorum vaginatum*. Generally thinly distributed except on flats of peat, where it is common.

Wild thyme, *Thymus serpyllum*. Common, especially in Strath Beinn Dearg.

Sorrel, *Rumex acetosa*. Occasional general occurrence.

Club mosses:
    Lycopodium clavatum
    Lycopodium annotimum
    Lycopodium nundatum?
    Lycopodium selago
    Lycopodium alpimum

Other mosses:
    Sphagnum spp, Polytrichum spp, &c.

Lichens: *Cladonia spp, Cetraria spp*.

} GENERALLY DISTRIBUTED, AND ENJOYED BY DEER

Plants which would be eaten if more common.

Blaeberry, *Vaccinium myrtillus* and *uliginosum*. Thinly distributed in dry places out of reach of deer. Occurs generally in Glac Cheann but is grazed to the peat.

Crowberry, *Empetrum nigrum*. Distribution as blaeberry.

Plants obviously present but not normally eaten by deer.

Bog asphodel, *Narthecium ossifragum*. Common, especially on lower slopes of Torridonian Sandstone.

Sundew, *Drosera rotundifolia* and *longifolia*. Common.

Scabious, *Scabiosa arvensis*. Thinly but generally distributed; more common on gneiss.

Mountain cat's-ear, *Antennaria dioica*. Occasional among the rocks, especially on gneiss.

Butterwort, *Pinguicola vulgaris*. Generally distributed. Alpine form also found.

Orchises, *Orchis spp*. Generally distributed below 1,250 feet.

MOVEMENT: THE INFLUENCE OF INSECTS AND FOOD SUPPLY

Peat bog flats between 250–1,250 feet

Providing new growth early in year.
   Cotton sedge, *Eriophorum vaginatum* and *E. polystachion*. Most common species on these flats.
   Sedges, *Carex flava*, and others. This species very common.
   Mosses, *Lycopodium spp*.
   Lichens, *Cladonia and Ericea*.

Alpine associations on peat-free gravels above 2,000 feet

The herbage floor of peat-free alps on Torridonian Sandstone.
   Sheep's fescue, *Festuca ovina tenuifolia*. General on alps, and viviparous.
   Alpine poa, *Poa alpina vivipara*. Occurs on alps.
   Alpine tufted hair-grass, *Aira alpina*. Occasional.
   Mat grass, *Nardus stricta*. Small alpine form, isolated on sandstone terraces round An Teallach.
   Alpine foxtail, *Alopercurus alpina* ⎫
   Woodrushes, *Luzula spicata*   ⎬ THINLY DISTRIBUTED
             Luzula arcuata   ⎪
   Highland rush, *Juncus Trifidus*   ⎪
   Sedges, *Carex curta*   ⎭
   Wild thyme, *Thymus serpyllum*. Reaches very high.
   Woolly fringe moss, *Rhacomitrium lanuginosum*. In large patches up to summits.
   Alpine chickweed, *Cherleria sedoides*. Amongst herbage of alps.
   Sorrel, *Rumex acetosella*. Thinly distributed.
   Cushion moss, *Grimmia*. Grows on boulders and much liked by the deer.

Occasionally nibbled by deer.
> Thrift, *Armeria vulgaris*
> Sedum, *Sedum anglicum*
> Purple saxifrage, *Saxifraga oppositifolia*
> Alchemil, *Alchemilla alpina*
> Cushion plant, *Silene acaulis*
> Epilobe, *Epilobium alpinum*
> Cloudberry, *Rubus chamaemorus*
> Cyphel, *Cherleria sedoides*
> Alpine pearlwort, *Sagina linnaei*

THINLY DISTRIBUTED WHERE NOTHING ELSE GROWS IN GRAVELS ROUND SUMMITS AND ON MEALL BHUIDHE

Other plants obviously present.
> Wild azalea, *Loiseleuria procumbens*. Occasional.
> Roseroot, *Sedum rhodiola*. In rock clefts.
> Melampyre, *Melampyrum sylvaticum*. Lower reaches of alpine region, and in woods of Gleann Chaorachain.
> Dwarf juniper, *Juniperis nana*. On peat-free gravels. Appears to be dying out.

The middle height, 300–1,750 feet, provides the bulk of the food supply of the deer. The very low ground is used in emergency only, and the peat-free tops, particularly of the Torridonian Sandstone, provide a special type of herbage which is grazed for the most part during the months of July, August, September, part of October, and during part of November.

In the compilation of the lists in this chapter I have used the nomenclature of Bentham and Hooker (1887), and the *Illustrations* (Fitch and Smith, 1887) to the British Flora have made these two works invaluable. *Types of British Vegetation* under the editorship of Tansley (1911) has also been used, but this work does not mention the distinctive vegetative complex of the West Highlands.

NOTE

There are, of course, no large predators in Scotland to affect the movement of adult deer, and it has been a constant tradition of the forests to avoid disturbing the herds as far as possible. It was a tradition I accepted until I watched deer of several kinds in the United States, and later the barren ground caribou in Arctic Alaska. The presence of the wolf keeps the deer moving; where there are no wolves, white-tailed deer and wapiti will both stay too long on one place and seriously overbrowse it. Movement by predators avoids this 'yarding' habit.

There is also the subject of helminthic parasitism in grazing animals which stay around one place too long. The hill sheep farmer in Scotland keeps this in check by means of an artificial wolf in the shape of a collie dog. The shepherd goes up the glens each evening with his driving dog which puts the sheep off the greens by the river and up to the middle slopes. The wolf must do the same thing for the deer where the wolf still exists. See Note on p79.

CHAPTER NINE

Reproduction

IN ORDER TO PREVENT tiresome subdivision this chapter will not directly continue the scheme of the last two as a study of reproduction in relation to movement. The aim, rather, will be to treat the subject of reproduction in red deer as a whole, as an activity in which movement plays a large part. It is at once a most interesting and disappointing chapter to write, because of the variety of the problems raised and left unsolved.

The emergence of sociality in animal life, in contra-distinction to mere associations of individuals, is closely bound up with the processes of reproductive physiology, and we can take this statement as axiomatic. Sociality may and does develop beyond the limits of reproductive expediency, but its origins are in reproduction, and we may look in the future to research on the hormonic processes concerned to throw much light on the development and interpretation of social behaviour. It is necessary briefly to survey the relevant field of sex physiology before considering the detailed behaviour of red deer in relation to it.

The Sexual Psycho-Physiology of the Stag

We are faced at the outset with the phenomenon of seasonal activity of the reproductive organs. Some animals, particularly those which have been domesticated by man or those which live in a modified environment, are sexually active throughout the year. The rat, mouse, dog, and domesticated cattle are examples. Others, on the other hand, are strictly seasonal. The female has one or two oestrous periods and the male reproductive apparatus undergoes considerable change for a sharply defined and almost constant period of the year. Red deer, squirrels, and voles are examples of this type. The external and internal environmental control and operation of seasonal sexuality is one of the major problems in sex-biology. It must be realized that testes and ovary are not autonomous endocrine systems. Their function is regulated from other sources which are the other endocrines, especially the pituitary body, the nervous system, and more general organismic influences (Lillie, 1932).

The gonads, also, have two functions – the secretion of sex hormone and the elaboration of egg and sperm. In the male the secretion of testicular hormone is the primary function of the gonad and it reinforces the secondary sexual characteristics of physical form, hair growth, voice, and temperament. The spermatogenetic function is attained at puberty and, in such animals as the red deer, for a short period of each year. For the rest of the year male animals of such species are in effect physiological castrates. What inherent physiological activity is at work producing these seasonal changes? The hypophysis or pituitary body certainly activates the testis to its function, but there we are foiled once more.

Rowan (1926–31) investigating the impulses to migration in the Canadian crow (*Corvus brachyrhynchos*) found that light had a considerable effect on gonadic development. By subjecting captive crows to long hours of light during the early winter months he was able to bring their gonads into spermatogenetic activity in the depths of winter, and when he liberated the birds they flew *northwards*, i.e. in the direction they would normally take in spring. He was also able to alter the sexual rhythm in the junco (*Junco hyemalis connectens*) by giving extra light in winter time. Bissonnette (1929–31) followed up this work on the American starling and showed that the gonad was developed by way of the pituitary body. It was the pituitary which was stimulated by light, and Bissonnette believes that light stimulation occurs through the eye. He has conducted similar light experiments with ferrets and the same mechanism appears to be at work. Baker (1932) concludes that the periodic development of the testes of voles (*Microtus agrestis*) is referable to the same cause. Incidentally, it appears that the infra-red end of the spectrum is concerned also with hypophyseal stimulation and not, as one might at first imagine, the ultra-violet end only. Bissonnette also found that activity had similar effects to light on his starlings. There can be no doubt of the light-activity basis of hypophyseal stimulation in the species studied, but it cannot be demonstrated to apply to all species.

Let us take for example the gonadic development of the woodchuck (*Marmota monax*) as studied by Rasmussen (1917). This little animal hibernates from November to March. The testes are at their minimal size and situated in the abdomen during August, September, and October, i.e. after the period of greatest light and activity. During the early hibernating period of November and December the testes begin to increase slightly in size, referable to increased spermatogenetic activity. In March, when the animal emerges from hibernation, i.e. after a period of darkness and muscular inactivity, the testes are ready for active spermatogenesis. (It is well known that in many mammalian forms it is necessary for the testes to be at a temperature some degrees lower than body-heat if they are to elaborate spermatozoa, a subject to which we shall return later. Hibernating animals experience a considerable drop in body-temperature. I suggest, therefore, that the testes are probably in an environment favourable to the development of their spermatogenetic function although they are still within the abdomen.) The testes of the woodchuck enlarge rapidly when the animal emerges from hibernation and are in the scrotum at the end of March or beginning of April. This is the period of sexual activity. Late in April the germinal epithelium is reduced to a single layer of cells with a few remaining spermatozoa in a wide empty lumen. The testes regress in midsummer and reach their minimal size again in August. It is probable that this regulation is hypophyseal in origin, for the administration of fresh hypophysis material to seasonal breeders awakens suddenly the secretion of hormone by the testes, but the environmental influence on the pituitary has yet to be discovered.

We may reasonably suspect that a factor of the external environment is concerned in stimulation of the pituitary, when it is remembered that when animals are moved from the northern hemisphere to the southern their reproductive rhythm becomes readjusted. The stag, which is in rut in September and October in Scotland, is in rut in March and April in New Zealand, in which country Scottish and English deer have been acclimatized. The same fact is noticed in the sheep. It may be found ultimately that cold and light are interchangeable factors acting on the gonads, light via the hypophysis and cold directly. This may be concerned with the fact that experimentally the administration of hypophyseal extracts (anterior pituitary hormone) brings about varying effects in different species of animals.

The Antlers

It is necessary to digress from the main theme of this chapter, for the sexual rhythm of the stag is complicated by the development and presence of antlers, and we are led into a maze of difficulties. Indeed, the stag presents a deep problem in sex physiology, and although I cannot hope to suggest solutions from pure observation, I shall attempt to set out the problem in some detail in the hope that it may be worked upon by physiological methods in the future.

Antlers are in themselves a problem. Of what use are they or what purpose do they serve? Let us rid ourselves of any ideas that they have a value as adornment or that what we look upon as beautiful antlers are of much use in fighting. They are not. About one stag in a 100 is without antlers and he is known as a 'hummel'. The hummel character, doubtless, is a genetic, sex-limited recessive, and perhaps this fairly numerous incidence, despite the fact that hummels are usually culled from the forests, occurs as a result of the relative success which hummels achieve as master stags. The genetic absence of antlers in no way affects the sexual rhythm or potency of the stag. Hummels have the reputation of being particularly wary and resourceful, they are usually in better physical condition than other stags, and they are able fighters. A hard dunt in the ribs from the polled head of a hummel seems to upset his opponent more than a sharp jab from the points of an antler. Were it possible to take a count of services by each stag in a large population I think it would be found that hummel stags would have covered individually a larger number of hinds than each of their antlered fellows. Their genetic significance in the population would be found to be greater than their actual proportion of incidence of one in a 100 would imply. Lack of antlers would appear to be a biological advantage to Scottish red deer.

Another common abnormality of the antlers is that known as 'switch-horned', in which the antlers show no other points but the brows and the terminal ones. It is generally accepted, and my own observations agree, that 'switches' are able fighters, and therefore capable in getting and keeping harems. There can be little doubt that numerously pointed and heavy antlers are a handicap. Their annual regrowth on lime-deficient soils is a serious drain on the resources of the organism. Antlers remain a puzzle. After all, we are not called upon to emulate the Swiss Family Robinson in finding a use for everything, but it would be interesting to know why and how they developed.

Huxley (1930) has suggested that in evolution antlers may have increased in size and branching with increase in size of the deer, and so, given the evolutionary value of an increase in body size, the elaborate antlers would not require to be of an adaptive quality in order to be controlled by natural selection. They may be merely a by-product of size.

Antlers do supply a means of defence for stags against predatory animals as well as of attack against members of their own species. It is possible that many-branched antlers would have a greater defensive value used against a carnivorous enemy. The adult deer have no such enemies in Scotland now, and it is altogether questionable whether the factor of attack by predators could be considered potent in natural selection for a sex-limited character in a species where the social system has developed along matriarchal lines.

The relation of antlers to the sexual physiology of the stag is beyond doubt. At one time I conceived the idea that the antlers served the purpose of a sexual valve in deflecting energy from the testes during the spring and mid-summer period. The hummel stags, however, experience no such deflexion of energy, and apart from their general good condition and the fact that mature hummels are usually among the earliest stags in rut, their sexual rhythm is unaffected.

Mature stags cast their antlers during the first week of April in my area, but each year I have noticed that the stags in the Beinn a' Chaisgein Beag area cast earlier than do those of Dundonnell. Good feeding conditions favour early shedding of the antlers. Almost immediately the new ones begin to grow and are covered by the nutritive, highly vascular skin called the 'velvet' (see p156). Growth goes forward steadily until August, when the antlers are complete. An abundant calcification of the antler at the coronet cuts off the blood-supply to the velvet, which then peels away from the antler as dead skin. The stags are extremely careful to preserve their antlers from injury during the period of growth. If they quarrel during that time, which is rare, they fight like hinds by standing on their hind legs and slapping with their forefeet. If they scratch their antlers they slowly lower the head and incline it to one side and use the hind foot with great delicacy. This is very different from their behaviour when the antler is free and hard. The velvet hangs from the new antlers in festoons as it is peeling off. I have seen the stags taking pieces of it in their mouths and gradually drawing it in, chewing it and thus helping to clear the antler. The stags of Western Highland forests waste nothing which can supply them with calcium phosphate, so they eat the stripping velvet from their own or their fellows' antlers. When the antlers are shed they are eaten from the points upwards and the butts only are left. I have even seen a hind chewing a stag's antlers while they were still on his head. By the middle of June there are no cast antlers to be found on the forests over which I have worked. This state of affairs does not occur so markedly in forests on better ground. I visited a forest on the eastern schist in central Ross-shire in December and found cast antlers still lying about the ground.

Until the blood supply to the velvet is stopped, the stags have short summer coats and the hair on the neck is short also. Thereafter, the mane grows strongly and is well developed by the time the rut breaks in the last week of September.

The antlers take on their full significance during the period of the rut. They are not used by the stag immediately they are clear of velvet, but as the rut approaches they are flourished in a neighbour's ribs

above: A rubbing tree
below: A young stag in 'velvet', June 1934

in playful manner. The observer may be surprised at the finesse and verve with which the antlers are used, for it must be remembered that fighting with the antlers is an activity shunned for the greater part of the year.

The antlers are used in another way during the rutting season. They form an erotic zone. The sexual activity of the stag is so intense that he has not sufficient hinds to satisfy his lust. He may masturbate several times during the day. I have seen a stag do this three times in a morning at approximately hourly intervals, even when he has had a harem of hinds. This act is accomplished by lowering the head and gently drawing the tips of the antlers to and fro through the herbage. Erection and extrusion of the penis from the sheath follow in five to seven seconds. There is but little protrusion and retraction of the penis and no oscillating movements of the pelvis. Ejaculation follows about five seconds after the penis is erected, so that the whole act takes 10 to 15 seconds. These antlers, used now so delicately, may within a few minutes be used with all the body's force behind them to clash with the antlers of another stag. These mysterious organs are a paradox; at one moment exquisitely sensitive, they can be apparently without feeling the next.

Behaviour of the Stag from the Casting of the Antlers until the Onset of the Rut

The stags are in companies at the time when the antlers are cast in April. They are peaceable and completely asexual. The testes are within the scrotum throughout the year, but are of small size between November and August. The stags remain in companies until the velvet is shed. At this time, a tendency for considerable movement among the stags is apparent. It is not altogether general, but new animals are noticed on the grounds; others which have been there all summer disappear. But there is no hint yet of the stags being in rut. I have mentioned already an Inverbroom stag being seen on Beinn Dearg Mhor, 15 miles away, in August; and the stag with the remarkably fine head which I saw in a company in Fisherfield in

September 1934, a beast which was certainly never bred in Western forests. The Duke of Portland (1933) records an interesting example of a travelling stag, presumably before the break of the rut. This animal had three antlers and his appearance was well known. He usually stayed at Langwell until August of each year. In 1873 the stag was shot at in the Reay Deer Forest, Sutherland, on or about 15 August, over 50 miles from Langwell. On or about 29 August he was shot at here in Dundonnell, more than 50 miles from Reay. He was killed at Langwell on 5 September. Langwell is quite 75 miles across the hills from Dundonnell. It is, in fact, a traverse from the west coast to the east.

A stag travelling in August and early September proceeds at a walk for the most part, but as the rut approaches his gait changes to a trot. This trot is no high-stepping, prancing gait, but one in which the legs are flexed as little as possible and each step is a comparatively small one. It is an economical gait.

During the week or two before the rut breaks, the stags appear to flock more than they have done for most of the summer. The bands move from end to end of their territories in the day and the observer cannot predict where they will be with the same assurance which he might have used a month before. At this time the stags are taking sly digs at each other with their antlers and the whole band occasionally indulges in a romp. This flocking of the stags just before the rut is difficult to assess accurately, because chance is an obvious element concerned with such an observation. I have noticed it pronouncedly myself and it is a general opinion among stalkers.

The stag's neck begins to swell in September so that when the rut breaks the increased girth of the neck and the mane of hair give the stag what is to us a most impressive appearance. Hingston (1933) has suggested the value of the mane and a large frontal area as a psychological weapon in fight, and there can be little doubt that such a purpose is served by this appearance of the stag.

In the week or two before the rut, the stags open up their wallows and roll in them once or twice each day, usually in the early morning or late evening. The mature stags wallow first and wallowing is continued throughout the rutting season. It is possible in mid-October to recognize the stags in rut and those which are not, from a considerable distance. The rutting stags are dark brown or almost black from wallowing in the peat, and the immature and late beasts still show the light red coat of summer. This autumn wallowing of the stags presents a problem. If the hinds wallowed as well at this time it might be thought that the emergence of the deer keds was the reason; but as the keds alight on both sexes indiscriminately, such a possible explanation loses all point. I think autumn wallowing by the stags will be found to be concerned largely with the sex physiology of the stag. Hingston (op. cit.) emphasizes the psychological value of the black-brown hair tufts or mane of mammals in fighting. The rutting stag is black-brown after wallowing, and as I shall point out in another chapter, deer are more frightened of dark objects than of light ones. Wallowing in early morning and late evening must produce a chilling of the skin, especially as evaporation progresses, and as the testes are then separated from the outer air merely by the thin skin of the scrotum, it is probable that they undergo direct cooling in this way. Stags take to the water much more readily at rutting time and will swim considerable distances in crossing inland lochs and arms of the sea.

The larynx of the stag develops during this period and he uses his roaring voice when he breaks into rut. The roar loses its volume and timbre as the rut decreases in intensity. The infra-orbital glands become active just before the rut and secrete a musky, amber-coloured, waxy fluid which trickles down the face and dries on the hair.

It is a question whether the secretion of testis hormone lowers the threshold of the tissues to pain. I am inclined to think it must do so. When the stags fight they do not seem to stop because of pain but from sheer exhaustion. Sometimes they will stop after a quarter or half an hour, rest a few minutes, and start again. In a fight or immediately afterwards there is never that sharp retraction of any part which may have been struck such as one sees when an animal is hurt in the ordinary way. Their hurts in fight seem to be of small account and a fatality is rare. The incidence of fighting is much exaggerated in popular literature describing deer. There is more noise and show than real fight.

20 September is the traditional date for the rut to break in this area of the Scottish forests. Nutritional conditions during the summer make for some latitude. The rut broke on 28 September 1934 and that was on Beinn a' Chaisgein Beag, the best of the ground. It was 1 October before roaring was heard in the Dundonnell forest. In 1934, after a late winter, the stags were a fortnight or three weeks late in casting their antlers. The early summer was dry when the grass was growing and the high hills were comparatively bare all the season. August and September were wet and there were few sharp diurnal alternations of temperature. The rut broke on 23 September 1935, after a year of good grass. But once again September was wet, nearly nine inches of rain falling in the month, and the temperature was steadier than if the weather had been finer with a few ground frosts during the nights.

We have still the problem of the hormonic linkage between the antlers and the gonads. It is known that the pituitary body is concerned in the processes of calcium metabolism. The growth of antlers is a phenomenon of calcium metabolism, and the annual casting and regrowth takes place at a time when hypophyseal (pituitary) activation of the gonads through the stimulus of light is taking place in some other animals. It could be postulated that the anterior pituitary hormone is being directed towards antler growth during summer and, as we know that no maturation of the gonads takes place until after the velvet is shed, that when the antlers are full-grown the hormone becomes deflected wholly towards activation of the gonads. Thereafter, the elaboration of testis hormone (which in our present state of knowledge we cannot attribute to any particular tissue of the testis) takes place, and the visible body changes in the stag follow. FIG 12 is an attempt to show diagrammatically the course of antler growth and the approach of the rut. Our hypotheses are purely tentative and only physiological experimentation can assist us to solve a problem so closely linked with behaviour.

If the gonads are damaged or removed during the course of regrowth of the antlers, growth stops at that time and the velvet is never shed afterwards in the animal's life. Does this indicate that there is a reciprocal action on the part of the gonads concerned

FIG 12

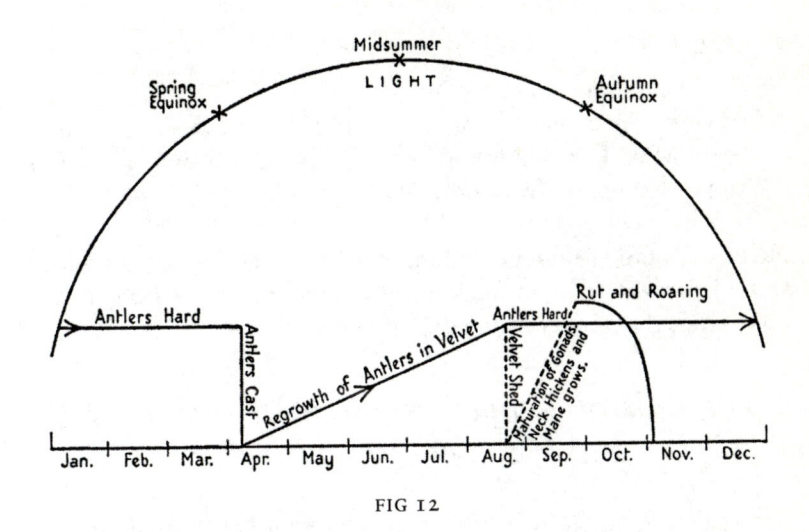

with the shedding of the velvet? Experiment alone can show. If a stag is wounded in August he does not come into rut at all, and if the injury is considerable on one side of the body, his growth of antler in the following year will be defective on the opposite side. Injury of one testis produces a similar result.

Antlers are a sex-limited character in red deer, but in the reindeer both sexes are antlered. It is worth while comparing shedding dates of antlers in this species in relation to reproductive processes. The data are taken from Hadwen and Palmer (1922):

1 MALES

Fawns. Shed antlers about 15 April.
Coining two-year-olds. Shed antlers shortly before females shed theirs.
Adult Bucks. Shed antlers after the rutting season in November.
Rutting season 25 August–October. Velvet peels 31 July onwards.

2 CASTRATES

If castrated in August when antlers are hard, antlers drop off in two or three weeks time. When castrated in velvet, antlers remain in velvet until spring.

3 FEMALES

Fawns. Shed antlers about 5 April.
Adult Does. Shed antlers a few days after birth of fawn in April or early May. If a doe fawns out of season, shedding of antlers similarly departs from normality.

Here is a totally different rhythm in another cervine species which should prevent us from making any hasty generalization on the relation of the antlers to the sexual cycle.

The Role of Temperature in the Development and Maintenance of the Rut

Before describing individual rutting behaviour I wish to discuss the possible significance of cold in relation to it. I have stated that 20 September is the traditional rutting date and that each year during the period of my work it has been later than that, following a wet and muggy August and September. Farther inland there were ground frosts at night each year at the beginning of September and the rut broke earlier. The traditional rutting date in Speyside (central) forests is 12 September. September frosts are the rule in that area and there is a greater daily range of temperature than in western forests.

We know from the work of Moore (1926) and Crew (1927) that the mammalian testis can function spermatogenetically only when it is at a lower temperature than the interior of the body. The anatomy of the scrotum is well adapted for the purpose of allowing the testes to be cooled from without. The skin is thin, in the stag covered with short white hair, but naked in the stallion and bull, and plentifully supplied with sweat glands. Sweating and evaporation produce chilling of the scrotum. The skin of the scrotum is very elastic, and although the permanent residence of the testes in the scrotum is ensured in the stag and many other of the higher mammals by a partial closure of the inguinal canal, there is considerable scope for their vertical movement. Warmth causes relaxation of the dartos muscle and the testicles descend. Cold causes contraction of the dartos and of the scrotal skin and the testicles are held nearer the body.

Many sheep farmers believe that sharp alternations of temperature – warm days and clear frosty nights – are stimulative of oestrus in the ewe. I have no means of demonstrating that a similar state of affairs obtains in the hind. There is, however, a closely comparable finding to record in the stag. Cold, dry nights mean intense sexual activity. The glens are full of the sound of roaring stags and, as the roar is protracted and jerky if the stag roars as he runs, the observer on a dark hillside knows that the stags are running frequently. If the following day is warm and dry the stags remain active. Now when clouds and mist cover the sky and the upper reaches of the hills and the air is humid, there is very little roaring to be heard and on such days the stags are visibly not so active with their harems. There is not the same amount of roaring, running, fighting, smelling at the hinds, or general restlessness as on days and nights of dry cold. I have noticed, also, that those stags with harems situated on the higher ground are more active than their fellows lower down. The ones I have in mind are those which were in the Polain, 1,750–2,250 feet (NE of Meall Bhuidhe), Garbh Coire Mor, 1,750–2,250 feet (NW of Meall Bhuidhe), and on the pass between Sail Mhor and Ruigh Mheallan, 1,750 feet.

Although the whole period of the rut may be six or seven weeks, each stag is not continuously active for that time. Those animals which come into rut first are the first to finish, their active period being about a month, but they are not active with hinds throughout that period. They may stay a week with the hinds and then be almost spent. They leave the hinds voluntarily or are driven out by new stags. Their behaviour in these short periods of two or three days of sexual inactivity is interesting. The stags go uphill into the cold, and the tops can be very cold in October. An observer may find no other deer but adult stags on the high ground at this time and they will be grazing and resting. The stags are not in herds, though there may be several on a square mile of ground, and no jealousy or pugnacity is displayed. The tops become a neutral territory. Heape, in his *Emigration, Migration and Nomadism* (1931), illustrates the concept of neutral territories by many examples from African ungulates. Whether in this instance the tops are tacitly accepted as neutral territories or the inclination to pugnacity has subsided with the lull in sexual activity is a subject for argument. My own opinion is against acceptance a *priori* of the tops as neutral ground, understood as such by the deer, although they may serve that purpose in effect. The cold appears to have a recuperating influence on the stags and at the end of the rut they collect in companies again on the high ground where they stay until December. If the weather of November is good they come down to the low ground afterwards in good condition, but if it is bad and they are unable to make the migration to the tops, they are in poor condition all winter.

When each stag is fit once more after the few days alone on the high ground, he descends to the breeding grounds and then only does he become pugnacious again. He molests none of his fellows on the tops and his passage is unhindered. The stags do not use their voices when alone on the tops in these rest periods during the rut, but as soon as they get down within the contour where hinds are likely to be met, or travelling stags moving between the breeding areas, they begin to roar. When they reach the rutting grounds the whole tempo of their movements quickens and a stag fresh from the tops has little trouble in chasing out one which has been among the hinds some time.

Movement of Stags during the Rut

Travelling stags at the time of the rut can cover considerable distances in short periods. Their pace is about six miles an hour, and that over rough ground is very fast. The stags often travel at night and rest during the day and distances of 10 or 20 miles may be covered at a stretch. These movements are exceedingly difficult for an observer to watch individually, for the handicaps of pace, light, weather, and keeping out of scent and sight are too great.

A fortunate combination of circumstances allowed me to follow six or seven miles of the movement of a running stag on 16 October

1935. I was on the western shoulder of Mheall Bhuidhe looking over Coire Mor at 11 o'clock in the morning. An eight-pointer stag which I did not recognize came into sight from between Sgurr Ruadh and Ruigh Mheallain. He trotted across the corrie, over the water, and up into Garbh Coire Beag out of my sight, never stopping or looking about. I climbed back up the shoulder and looked over into Garbh Coire Beag, to see him climbing out of the corrie. He crossed over the rounded top of Meall Bhuidhe and down into Coir' a' Mhuillin where I picked him up again. There is a pass over Glas Mheall Mor between Coir' a' Mhuillin and the Glas Thuill corrie, near to where Glas Mheall Mor joins the main pyramid of An Teallach. The stag climbed slowly over this and I, who was still on Meall Bhuidhe, lost sight of him again. The ground is easy on this hill and I was not long in skirting the head of the corrie and reaching the pass where he had gone down into the Glas Thuill. I could see him trotting down the long and well defined deer path which leads to the lip of the corrie. It was easy for me then to follow the ridge to the summit of Glas Mheall Mor and pick up the stag again. He trotted in and out of the terraces of the Torridonian, and I eventually lost sight of him going down into the pine wood. This trek had taken him an hour and a half. His route was not the best he might have taken, which would have been to come farther up Coire Mor to miss Meall Bhuidhe and to reach the pass on Glas Mheall Mor without dropping into Coir' a' Mhuillin.

When a stag is in rut he eats very little except mosses and he is, of course, very much more active physically than during the rest of the year. His form undergoes considerable change, in that his belly runs up until it is as small as that of a racehorse and sloping upwards to the fork of the hind limbs. The swollen neck, the mane, and the shrunk hind-quarters give the stag that appearance which Hingston emphasizes in his paper on the psychological weapons in animal fight. He appears to have an exaggerated fore end and an attenuated rear. A stag in August, on the other hand, has a big belly full of lush food and his neck is comparatively thin.

The Course of the Rut

The social impacts of the rutting season are intricate and various and the observer deplores the lengthening nights which make for discontinuity in his work. This is all the more annoying when so much behaviour of importance takes place under cover of darkness. Stags move more frequently then; I have heard the clash of antlers in the night, and as it is extremely rare to witness the sexual act in Highland deer, it is reasonable to assume that this usually occurs during the night also.

I heard roaring in the 1934 season on 28 September. That was in the north-eastern corrie of Beinn a' Chaisgein Beag at an altitude of 1,750–2,000 feet. There were 77 hinds and followers in the corrie and there was one big, dark-coloured stag with them who was easily recognizable by his particularly wide spread of antler. These deer were grazing as a group on a space of about 20 acres and the big stag was running round them continuously, roaring every minute or less, scraping up the ground vigorously with his forefeet, lying down for 30 seconds, up and running round the group again with his muzzle outstretched and roaring as he ran. He masturbated twice within an hour and not once during the whole afternoon did he approach a hind to mount her. Two or three hundred yards away on each side of the group were some young stags and a few rutting stags. The youngsters took no notice of the central group of hinds and either grazed or lay quietly. The rutting stags were in no way equal to the big fellow who was with the harem, and although they stood and bellowed they made no effort to challenge his superiority. If they came a little too near, within 200 yards of the hinds, the big stag would run out towards them at a swift trot, his muzzle far extended. They never stayed to meet him and he would not chase them far enough to lose touch with his very large group of hinds. The females were quite impassive to the noise and activity which was going on round them. They continued to graze or lie down chewing cud.

Two days later 'Wide-spread' was still in the upper part of the corrie on Beinn a' Chaisgein Beag, but he had only 46 hinds and followers with him. There were three stags with harems in some broken ground below the corrie and they had 10 or 12 hinds each. The rutting territories of these three stags were much smaller than that of 'Wide-spread' higher up, and they were closer to each other than to him.

On 4 October the stags of Beinn a' Chaisgein Beag had taken still more hinds and 'Wide-spread' had only 23. The others were in the hollow below, about half a mile above the Uisge Toll a' Mhadaidh at 1,250 feet. 'Wide-spread' had only 11 hinds and followers on 7 October and he was not nearly so active as on previous days. Then he disappeared. By this time the volume and persistence of roaring had increased among the other stags and Beinn a' Chaisgein Beag was a place of great activity. There were 10 harems on the north-eastern face of the hill. Twelve more rutting stags were in the vicinity, standing on the far side of knolls and looking down at the harems. Their positions were such that only their heads were showing from where the harems were, but as they were roaring from time to time their positions did not conceal their presence from the stags, with the harems. The activity on this side of the glen was in striking contrast to the quietness on the opposite face. Here there were only the few hinds of what I consider to be the new Creag-mheall Meadhonach hind territory, and there was one old and not very active stag with them. Map 3 shows clearly how the rutting territories are bunched in particular areas.

During the season of 1935 there were two or three of what I am reasonably sure were the same stags as were present in 1934, in the breeding ground of the north-eastern corrie of Beinn a' Chaisgein Beag. There may have been more than that, but I know some were shot and others were not sufficiently distinctive. 'Wide-spread' did not appear again in 1935, and I am afraid his head had too great a trophy value to give him chance to survive.

The north end of Carn na Carnach is of great interest at rutting time as several harems congregate there. At this place I have watched many examples of behaviour of rutting stags towards each other on the boundaries of their harem grounds, and of the way in which only stags of almost equal merit fight each other. Here are a few extracts from my notes taken at the times mentioned:

1934
5 October

An old switch-horned stag has had 15–20 hinds on Achachie for several days. Now – 5pm summer time – the weather was still rough and wet and temperature was falling, 50–45° F. Six young stags were with 12 hinds on Achachie but no sexual activity being displayed. The old switch came over the hill from the western side of the Carn with six hinds. He ran fast to the group of 12 and with unerring selection and purposeful intent he chased out of the group each young stag, one by one, until they were 150–200 yards away. None of them stopped to contest the point. Then the old switch added the 12 to his six at great speed and stopped to roar. He kept herding them by running round and round the group in great excitement. Sometimes he ran clockwise, sometimes the other way. He began to work the hinds back over the north shoulder of the Carn out of my sight but left four on Achachie. Over in the hollow (marked 1 on FIG. 3) were two more groups of hinds, nine and six, and two eight-pointers were on the slopes north and south of these hinds. The stags approached the two groups of hinds which then joined into one group. These two stags, which were fairly fresh and in fine fettle, took good care to keep out of the way of the old switch and he took no notice of them. Now the two approached each other, walked round one another at 10 yards range, and roared. Their heads lowered, the hind quarters were depressed, and the hind feet were farther backwards from the body than normally. The stags took slow steps forward, then a quick run, and their antlers were together. But they were not yet fighting. These rapid passes of the tips of the antlers remind one irresistibly of the preliminary passes of two fencers. The feet were firmly planted and put no force behind the movements of the head and antlers. And now one of them side-stepped and lunged, but the other wheeled his hind end and met his opponent's antlers with his own. The point in all these fights is to lunge a broadside in the opponent's ribs, and to prevent this happening to one's self. Much of the fight, therefore, consists of a shoving match, antlers to antlers. The stag which can shove the harder is usually the winner, his opponent retiring before he gets a broadside. For in this shoving there is a distinct tendency to push not quite head-on, though the opponent tries to keep the shove directly head-on as far he is concerned. If the shove can be strong enough and at a slight angle, the other stag is forced from the straight and a quick jab can be taken at his ribs or belly. One of these eight-pointers got such a blow after five minutes and he retired. The winner roared, herded his hinds, and roared again.

At this time another old stag with poor antlers came up to Achachie from the Gleann Chaorachain burn, roaring as he walked. He took possession of the four hinds left by the switch who had gone over the shoulder and began to herd them. Not five minutes had passed, however, before 'Old Switch' came back over the hill at a fast trot, his head low. The new stag did not stop with the hinds, but made off at a gallop towards the quartzite slabs below the Carn. 'Old Switch' did not follow him but went round the four hinds at great speed and chased them over the hill, presumably to the others.

It was now almost dark and I could see no more. But in an hour and a half on a small space of ground there had been considerable change in the constitution of the harems and a succession of social impacts. Such is the course of the rut.

16 October 'Old Switch', the eight-pointer, and the old stag mentioned last night had harems on Achachie. The first two, although not attempting to fight or take each other's harems, were constantly challenging each other along their boundary. They would approach to 10 yards' range, roar, and then walk side by side, but getting nearer. One would try a quick jab at the other, but as he was ready for such a move and had side-stepped and lowered his head in readiness, all in half a second, the first stag would not complete his rush but wheel parallel to the boundary again and the walk would continue. They would turn, walk back along the 100 yards of boundary, one or the other attempting a jab but never achieving it or following up towards a fight.

On this day I saw all the deer on Achachie throw up their heads and, jealousies forgotten, they ran away in a bunch over the north shoulder of the Carn. Later on I followed them and they went far over Coir' a' Ghiubhsachain. The cause of the trouble was a collie dog (a stranger to this country, for I know all there are) who came over the south-eastern part of Carn na Carnach, chasing a six-quarters staggie. They disappeared in the trees by the burn at the foot of Achachie. Later in the day, just before dark, the Carn was still clear of hinds, but the eight-pointer which was beaten last night was roaring by himself on Achachie. I saw him walk in a straight line towards a small bed of brackens near the northern edge of Achachie and as he seemed to have an objective in view I looked ahead along his line and saw a beast which I am almost sure was the six-quarters staggie which had been the quarry of the collie dog in the morning. He looked unhappy, for his head was low, he was not grazing, his back was arched and his legs hunched under him. The bigger stag approached him, lowered his head, depressed his croup and gurgled a roar deep in his throat. The little fellow looked round pathetically, as it seemed to me, but he did not move. One terrific dig, and one only, in the inguinal region from the eight-pointer bowled him over. This stag then turned in his tracks and walked back the 300 yards to his old position in as straight a line and as slowly as he came. The little beast got up, hobbled a few steps with very stiff legs, and stood as he was before the incident occurred. Darkness came, and I could see no more of him, but a week later I found him dead in the long heather just below Achachie, not a hundred yards from where he was hurt.

On 17 October the main bunch of hinds had not returned to Achachie. There was the eight-pointer of yesterday and one hind. The temperature rose 6° to 56° F and sexual activity waned.

It has been mentioned that only stags of equal merit fight each other. No stag will face another better than himself. What enables the stags to assess each other's form? We do not know. There are ample physiological reasons to account for the stags coming to fighting condition, but beyond that a mechanistic explanation cannot go.

The sense of smell is the most highly developed in deer and during the rutting season it becomes obvious to the observer how important this sense is to the stag. He is constantly licking his muzzle when with the hinds, raising it into the air and retracting it from the upper gum. This is a posture common to many animals in the first stages of sexual excitement, especially ungulates. The sense of smell is always extremely selective and its operation at long range is interesting to watch. I have recounted how 'Old Switch' came back over the north shoulder of Carn na Carnach when another stag took possession of four hinds he had apparently left behind. Here is another incident which I think worthy of record.

On 20 October 1934, I came over the pass between Ruigh Mheallain and Sgurr Ruadh to watch some stags with the Gruinard hinds on the steep slopes above Loch an Eich Dhuibh. The wind was from the south-west, and as these slopes are scored by gullies draining to the loch, it is comparatively easy to stalk close to deer there. As I crawled down one gully I could hear a stag roaring to the north-west of me. My way was blocked by a hind and her yearling stag and stag calf which were grazing lower down in my gully. I waited over an hour for them to move over to the group with the stag. They did not do so, and the stag, who was roaring hard most of the time, never came in sight. All this was wasted time to me, who wanted to observe the behaviour of this harem, and I decided to show sufficient of myself to move these three over the edge of the gully without actually frightening them. Hazardous as this procedure may appear on paper it is quite possible to manage successfully if care is taken not to go more than one move beyond the point of arousing curiosity. If you go only far enough in attracting a hind's attention that she comes to find out for herself, or you actually frighten her, the game is up. But there is just a point when she will move away slowly and not frighten others out of sight. In this instance, the hind did not actually see me, but the staggies did, and she and they went down into flattish ground at the south-east corner of Loch an Eich Dhuibh. As I suspected, and found out for myself another day, the rutting stag could not see the hind in the place she was now, 400 yards away. I crawled up to the edge of the gully and heard the stag running and roaring and coming closer to me. A big, strong-antlered, six-pointer came in sight a hundred feet below me. He trotted obliquely down the hill straight to the hind I had moved and he brought her back to the others at a fast trot, his head outstretched and his muzzle more than two or three feet behind her. The staggies walked up leisurely after them. Now, here was one hind who had not been with the stag or the harem for over an hour, as I knew from my long wait; but as soon as she moved away 400 yards, not into his field of vision, he came from his group directly towards her and ran her back to the others.

An observer may witness considerable sexual activity and herding behaviour on the part of the stag, but the act of copulation is rarely seen and I have remarked on the general impassivity of the hind. I have only twice seen a stag cover a hind, both times on Carn na Carnach – once in 1934 and again in 1935. On the first occasion at about 11am, the stag mounted the hind 12 times in the course of half an hour, but she ran away 11 times and only at the 12th did his hind feet leave the ground, a sure sign that the act has been completed. The second occasion was on 18 October 1935, at about 4 o'clock in the afternoon. The same stag as last year, 'Old Switch', was there with hinds. One was in season. The stag approached her; she ran away; he chased her (20 yards); she stopped and came to him; rubbed her whole length along his ribs from fore to hind end; made as if to mount him; he turned to mount her; she ran away; he chased; she stopped; he mounted and served her. When he was sliding off she kicked up her heels, hit him the belly, and ran away. He roared and followed her again; she came to him and rubbed herself along each side; licked his muzzle, walked under his chin, throwing back her head; licked his sheath for a moment; made as if to mount him; he turned, mounted, and served her again; roared; walked round the group; lay down for a quarter of an hour; rose and walked round the group; lay down again for half an hour. Another stag came over from beyond the Carn at 5 o'clock, but ran away as soon as 'Old Switch' rose to chase him.

This incident showed true courtship behaviour on the part of the female, but it is rarely seen by man. On 22 October I watched a one-antlered stag on Creag-mheall Meadhonach. He came up to a recumbent hind and touched her muzzle with his. She gently relaxed, lay full length on her side, stretched her legs, and her muzzle was extended along the ground. The movement was beautifully co-ordinated. The stag remained with his head lowered to her for a full minute, his muzzle slightly indrawn and the neck arched.

I have mentioned above that 'Old Switch' was back on the same ground in 1935 which he occupied in the rutting season of 1934.[1] His usual position was such that I could watch him and his harem easily, and his period with the hinds will be outlined as a concluding example of behaviour at this time.

---
[1] He was there, also in October 1936.

1935

13 October — 'Old Switch' came over the north point of the Carn at 4 o'clock in the afternoon and this was his first appearance this season. He came at a fast trot and roaring. An old stag, far run, was with some hinds on Achachie but he ran away as 'Old Switch' came trotting in. The hinds were indifferent to the change. He herded them at 20-minute intervals until dark.

14 October — Humid weather and little activity. 'Old Switch' herded his hinds at 20-minute intervals by walking round them. He approached a resting hind and they licked each other's muzzles for five minutes. Their lower jaws oscillated *vertically* (opposite to the transverse cudding movement) until a foam appeared round their muzzles.

15 October — The stag had been wallowing. There was little activity and the south wind blew a gale of great force.

17 October — Cold north-west wind and snow to 1,700 feet on An Teallach. The weather was very cold, very wet, and very gusty. The harems were broken up, the hinds grazing in threes and fours and the stags standing alone and silent. The whole picture was very striking. The deer were easy to approach.

18 October — Wind round to south and much warmer. Harems together again, but 'Old Switch' has got 19 hinds instead of his usual 11.

19 October — Wild west wind and two inches of rain. Harems disorganized.

20 October — A cold north wind; sprinkling of snow in the Strath, but it receded to the 1,000 foot level by the evening. The harems were gathering together again. (Cf. thermo-hygrograph records, FIG. 8, p. 112)

23–25 October	Better weather. There was fair activity and roaring on Carn na Carnach.
26 October	A day of continual rain and poor visibility. The deer were low and there was little or no activity.
27 October	Mist down to 300 feet. Heard 'Old Switch' roar twice during the morning, but there was obviously little activity. I could see nothing of the deer for the mist.
28 October	Wild weather and snow down to 250 feet. Sexual activity on the part of the stag slackening markedly.
29 October	Frequent hail-storms. The old stag had 48 deer close round him below the north point of the Carn. There were seven young stags on the outskirts, nine calves, 12 yearlings and two-year-olds, and 20 adult hinds. These deer were on their feet from three hours rest at 2pm and grazed southwards below the western face of the Carn. The leading hind was in front of the group and 'Old Switch' was in the middle, giving an occasional growl but showing no sexual activity. He did not object to the presence of the young stags, two of them six-pointers, which were mingling with the rest.
30 October	Another very bad day with much thunder and lightning. 'Old Switch' was wandering as far as 300 yards from the group in the main hollow on the north-east side of the Carn and making no effort to run.
31 October	The stag still with the hinds on Achachie, but his behaviour in no way differentiated him from the rest.
1–4 November	An improvement in the weather. Some sun and light winds from south and south-east. Hinds getting back into their own groups. 'Old Switch' disappeared on the 3rd.

On 6 November, which was a good day, I went a long round of the forests. The Carn hinds were on the top of their hill. Stags were in groups at 1,500–2,000 feet on Beinn Dearg Mhor and there were many hinds in their usual grouping about the saddle between Beinn Dearg Mhor and Beinn Dearg Bheag. Some young stags were roaring in Strath Beinn Dearg and on the lower slopes of Beinn Dearg Bheag, but they were not running the hinds. Some stags were with the Gruinard hinds above Loch an Eich Dhuibh but were not active. There were spent stags high in Coire Mor, on Meall Bhuidhe, and in Coir' a' Mhuillin, where the ground was frozen hard and the air was dry and cold. On coming down to the Dundonnell pine wood I heard roaring and in the moonlight saw some hinds and a stag with back-curving antlers. He was actively running. This beast was with the hinds in this place a month before. He disappeared about 20 October and I saw him in Coir' a' Mhuillin on 1 November. Now he was down again and active for another three days, when he disappeared once more and went to the tops.

Young stags may be heard roaring during November, but their behaviour contrasts with that of older rutting stags. Young beasts may be sexually excited and will serve a hind if opportunity offers, but they display none of that ability in herding a group of hinds together which is such a striking feature of the adult stag's activity. I have heard a three-year-old stag roar in a desultory fashion as late as 15 December, but he was then merely a grazing member of his mother's group and he was not sexually active.

Thus ends the period in which the stag plays his part in reproduction. There still remains something to be said of the behaviour of the hind at calving time.

The Hind at Calving Time

The hinds which are about to calve leave the group for a few days. Calving is very rarely witnessed by man, and it occurs presumably in the very early morning. I have heard the slight bellowing in the night-time which a hind makes when the calf is being born. I have never seen parturition in the hind, but I set down here an account I took on the day following the incident from Robert MacDonald, Achneigie, who saw a hind calve on Beinn a' Chlaidheimh, opposite his house.

At 9pm the hind was bellowing a little and walking round one spot in the heather. The following morning she was doing the same, lying a little, getting up, pacing about, and lying down again. At 3pm she dropped the calf and was standing at the time. She turned round and looked at the calf for two minutes without moving. Then she walked away a few paces, began to lick herself, cleansed and ate the cleansing. She came back to the calf, now lurching on to its knees, and looked at it again for five minutes. Only then did she begin to lick it. At 3.30pm the calf stood up for a few moments. Then it lay quiet and the hind walked away 100 yards and began to graze. There was very little maternal solicitude apparent. This is in line with my own observations on hinds and calves during the first three or four days of the calf's life. I have stated in an earlier chapter that the calf lies alone for the first few days and is visited by its dam about twice a day to be suckled.

When the calf is on its feet, however, a great change occurs. The calf suckles every few minutes and the hind fondles the calf very frequently. The hind moves slowly over the ground as she grazes, turning constantly to the calf which, even if it is not yet strong on its legs, can move over very rough ground by crawling or shuffling. She nuzzles and licks the calf, and the ears particularly are licked with vigour. This licking of the ears and back of the neck continues until the calf is three or four months old. If one hind starts doing it, the others will soon follow suit, and it is very amusing to see several calves standing like little boys having their ears washed by their mothers. The outstanding error in the analogy is that the calves appear to like the licking. I do not know why this is done. Ticks tend to gather on the fore end of animals and the lick of a hind's rough tongue might well tear them off, but I do not think the calves are much infested by ticks in these early days of their lives.

It is probable, considering the usual lactation curve as measured in cows, that the hind gives very little milk during the first three days after calving. If we keep in mind the findings of Turner (1934) that lactation is initiated and regulated by the secretion of anterior pituitary hormone, and of Wiesner and Sheard (1933) that maternal behaviour can be induced in the rat by administration of this hormone, we can postulate that an increased secretion of anterior pituitary hormone in the hind takes place within a few days of birth of the calf and that increased milk production and enhanced maternal behaviour are linked in this manner. I have already alluded to the long lactation period of the hind and its social significance. We may look upon the secretion of the anterior pituitary as the hormone which is the foundation of sociality.

NOTE

The late Duke of Bedford wrote to me on two occasions questioning the accuracy of my estimate of a long lactation period in the hind. His experience at Woburn and at Carsphairn was, he wrote, that hinds were dry by December. He suggested that the almost certain subsequent death of the calves of milk hinds shot in the Highlands would be referable to loss of mothering rather than of milk.

There are not many mothers of calves shot intentionally in the Highlands so investigation on this point has not been easy. Nevertheless, such few December milk hinds as I have examined have certainly had milk and many stalkers have supported this finding. I have known mares living out with no more adventitious food than hay to lactate from foal to foal, an interval of a year. The quantity of milk may be very small, and in the case of the mares, I know it was very small, but I believe it has a value for the young animal quite out of proportion to quantity.

CHAPTER TEN

The Senses

THE ANIMAL'S EXPERIENCE is its own and a closed book to us. We must be extremely careful not to obtrude our human view of things in our observation of the reactions of an animal to an environment, in so far as it perceives that environment through the senses. An animal's perception is in a large measure governed by the structure of its sense organs and by its power of movement. Thereafter, perception is modified by the needs of the animal, by its efforts to maintain an ecological norm. The patterns, configurations, or *gestalten* perceived in a single environment by several species of animal are widely different. And the patterns may be changed for one species and each individual, depending upon the significance or valency[1] of any part of the environment in the animal's attention at any one moment. The environment itself changes also for many animals in the course of the year, so that life for the animal which presumably does not contemplate is, in effect, a kaleidoscope of patterns or configurations (used in the broadest meaning of these terms to include the operation of all the senses) linked and centred in the organism by instinct, association, memory, and conation, and resulting in what we understand as the behaviour of the animal. In studying animal behaviour it is our task to find out as far as possible the nature of the animal's perceptual world, to discover the valencies of things in space and time. This boils down to getting a more thorough knowledge of the animal's environment, not purely for our own philosophical contemplation, but in order to simplify it and relate it objectively to the animal's needs and perception.

---

[1] This term has been suggested by Russell (1935) as an improvement on the terms 'significance' or 'meaningful stimulus', which can be given a teleological connotation. I use it here and elsewhere in connexion with things occupying the attention of deer.

I think that the fact of the animal's primordial curiosity should be constantly in our minds when observing or discussing its senses. It is this fundamental curiosity which goes far to ascertain relations and to attain comprehension. It has been shown in the course of this book how often inquisitiveness is brought into play, overriding instincts which might have been expected to prompt actions of a different kind.

The senses of red deer are acute and selective, and behaviour has therefore a vivid quality and wide range which enables us to make some estimation of the effect of several factors in the environment with greater facility than could be expected with animals of lesser sensibility, or of less latitude of reaction to, and movement within, the environment. At the same time, we must not over-estimate the value of observation of normal behaviour in gaining knowledge of the senses of animals. Experiment may provide short cuts and can elucidate problems which years of pure observation will leave unsolved. Observation is the first and necessary step in that it may prevent experiments being staged under conditions which include, and do not account for, serious abnormalities in the sensory environment.

Smell

This is by far the most meaningful sense in the lives of red deer. Their reliance on the information it conveys is absolute. When, for example, deer smell man, they do not wait for confirmation of the fact by the exercise of other senses; they move away. Localization of the source of the scent exciting attention is remarkably accurate. Wind is naturally of great importance in the conveyance of scent, but if the wind is of such unstable quality that it is a succession of eddies from all points, the deer may be driven to panic by inability to localize the source of a disturbing scent. An incident occurred on 14 April 1934, which illustrates this. I was descending the quartzite cliffs of Coir' a' Ghiubhsachain, the wind being fitful from the west and south and the humidity high. Two stags were on the other side of the corrie approximately half a mile away. Up went their heads, but they looked southwards and not at me. They wheeled with their noses towards An Teallach, back again, and over towards me. It was obvious that they did not know the direction or distance I was from them and I do not doubt it was my scent which had been carried to them on a most unreliable wind. They trotted uphill 100 yards and tried the wind again by raising their muzzles to all quarters. Still they could not place me, and, considering the dull light and the distance, it was unlikely that I could be seen by them. They turned towards the Toll Lochan corrie and went over the rough ground at the gallop, then over some quartzite slabs towards Sail Liath. At that time there was a large snowfield on the north-eastern face of Sail Liath, topped by a considerable cornice. The two stags went into the snow, plunging upwards as fast as the conditions allowed. The deer are in poor condition at this time and their frightened movement severely winded the two stags. Outlined as they were against the snow I could see through the glass the open mouths, the tongues prominent, and the breath coming fast. They plunged finally through the snow cornice, sinking to their bellies, and once on the hard ground at the top they went quickly out of sight on to the Strath na Sheallag side of the hill. These two stags were not strangers to me, although I had but then newly taken up the work, and it is unlikely they would have made such a long and difficult journey had I surprised them by scent and sight at 100 yards' range. Had they seen me moving steadily at the distance they were from me at first, it is probable they would not have moved at all.

The relations of humidity and temperature to the sense of smell have been referred to already in the chapter dealing with the meteorological influences on movement. Constant variations in humidity provide an olfactory stimulus to scent reception and perception and this effect is heightened at fairly high temperatures. Frost and the steady humidity frequently prevailing at the time appear to lessen olfactory reception and perception. Here is an account of the most remarkable incident of its kind which has occurred during my observations. On 5 November 1934, the weather was dead calm and cloudless above a good layer of snow. Just before dusk the temperature was 30°F and I was about to walk up Glen Chaorachain to watch the Carn deer which were down on the quartzite slabs. I did not get within 70 yards of the burn, for the deer were coming through the birch trees and over water. I stood against the trunk of a birch, motionless. The deer had not seen me and they came slowly on, browsing at tufts of bents and heather which showed through the snow. The light was failing, but the snow and clear air helped to keep objects sharply visible. Three hinds and a stag approached the trees under which I stood. They rose on their hind legs and browsed on the twigs of the very one above me. I could have touched them and their breath came in my face. Many more of the herd were within five to 20 yards. I had the feeling of having reached that state which all watchers of animals desire, of having dispensed with my physical presence. The deer moved on in their grazing in twos and threes, and I was left alone under my tree in the still and silent air. They seemed to have been unaware of me only three feet away from them.

When deer are uncertain of the evidence of the senses of sight and hearing they try to verify their suspicions by the exercise of the sense of smell. Let us take, for example, one incident occurring at close range of which the movement was typical. At longer range every observer of deer sees similar behaviour at one time or another. On 17 November 1935, in good weather with a slight south wind and just before dusk, I was creeping through the herbage from an easterly direction close up to a hind and calf in the Dundonnell pine wood. The hind heard the rustle of my approach and became suspicious. Her calf stood still, but she moved through an angle of 90 degrees to reach a position 25 yards north of me. This is where she should have got my wind, but I was low in the herbage and her head was held high, so she must have missed it. I could see only her expressive ears. They came forward as she was inquisitive; they went up a little and one went back as she was doubtful; then both went up straight and close together as she was annoyed. Then she barked between 40 and 50 times at five- to 10-second intervals. I did not move and still the scent did not reach her. She walked away, head in air and stiff-legged, back to her old position and had another round of barking. Still I did not move. The hind came back on her line to leeward of me. She came a little nearer this time but my scent evidently did not reach her. She walked back again to her calf which had not moved these 20 minutes, and together they moved away slowly to a distance of 100 yards. Lacking the verification of scent, but knowing by sight and hearing that some extraneous thing was there, she had not taken to flight. I mentioned on page 112 a young stag which at a distance of a few yards walked round me into my wind before making off.

The few hinds and followers which I fed in Glac Cheann behaved in an interesting manner from time to time in relation to my scent. It was a long time – a month – before they would approach and eat my food. Thereafter it was eaten up each night. Occasionally, I would walk to the head of Glac Cheann during the day or over the quartzite slabs at times when the hinds were over in Coir' a' Ghiubhsachain and not visible to me nor I to them. That night they would not touch the corn. I, as corn-bringer, could walk to the feeding-place to within 100 yards of them, in their field of sight and smell. But I was not allowed to go beyond this point. After that I was an object to be feared, whether seen directly, or smelt some hours afterwards. These deer appeared again in Glac Cheann after their summer on the high ground on 8 October 1935. The hollow could be observed a high degree of accuracy as it was directly opposite Brae House. I went to the old feeding-place and put corn down the following day, thinking I might manage to get some stags feeding too by time the winter came. The hinds ate the corn the first night it was down. The inimicality of my scent as far as the feeding-place was remembered, as well as the pleasant taste (and sight?) of the food. This was at a time, it should be remembered, when there was no great urge to accept adventitious food. I have remarked in Chapter Three how these hinds must have assiduously followed my tracks in the lower part of Glac Cheann, finding the handfuls of corn I put down here and there. Within one period of time but at different places in their territory, my scent was valent for these deer in opposite directions – on the one hand, as a thing to be sought and followed, on the other, as something of which to be frightened sufficiently to keep them away from the corn which was awaiting them in the lower part of the Glac.

The approach of snow which, we have seen, is sensed by deer, is probably perceived through the olfactory sense, for the drop in humidity and veering of the wind to the north appear to me as the common factors heralding snow.

For the stag the sense of smell is particularly significant at the rutting season, and incidents have been recounted in the previous chapter which illustrate this aspect of the olfactory sense.

Hearing

This sense is the next most highly developed in deer so far as observation can help us to decide. Hearing may be looked upon as the inquisitive sense among animals, for if the sound they hear is unknown to them and not of sufficient volume to frighten them immediately, they tend to approach it. The gamekeeper squeaks on the back of his hand to bring a weasel from a dry-stone dyke and, poor beast, the ruse seems never to fail. The poacher can squeak a hare near enough to shoot it with a catapult, cows will go a long way to hear a brass band, and dogs are intrigued immediately by strange sounds. Stags are not very inquisitive, but hinds are and they investigate the source of strange sounds. I have drawn them nearer by whistling, and even singing has not driven them away from 70 yards' range. And then, to note the effect, I have scraped the metal of my shoe toe on a stone and at once they have moved away. Their hearing seems to be highly selective. In the noise of a high wind they will hear an observer's approach within 50 yards by the rustle of heather or the squelch of his passage over the wet ground, even though he cannot hear those sounds made by himself.

I have noticed repeatedly, however, that deer – and particularly stags – will lie in 'zones of silence'. The observer may stalk to within 100 yards, using normal care. After that, approach is much more difficult and then, within 50 yards, he notices that quite suddenly the sound of the wind in the herbage has ceased. The slightest movement causes what seems an excessive amount of sound and the deer are aware.

With both this sense and that of sight, the deer associate certain events with danger. The sound or sight of grouse flying away in frightened fashion is sufficient to move deer at the trot without waiting to find out what caused the disturbance. Similarly, sheep in a deer forest are most upsetting to an observer. They are rarely shepherded and in some instances never, for some go practically feral. Such sheep appear to have more acute vision than the deer, or at least they can recognize the figure of a man in a greater variety of postures. These sheep give a sharp hiss or snort down their noses and away go the deer. If the deer are too far off to hear the hiss of the sheep they may see the animal's white shape, moving hurriedly, from a considerable distance, and that is enough to move them. The warning hiss of a wild goat has the same effect but it carries farther, being more sudden and explosive than that of the sheep. Fortunately, the goat does not run away and warn the deer so easily through the sense of sight.

Sight

The sense of sight in red deer is very interesting to the human observer, largely perhaps because it is his own most vivid sense and he can note their behaviour in relation to vision in a greater variety and complexity of situations than he can in relation to the other senses. The sense is acute, but what is most evident is its practised quality. The vision of deer is most probably inferior to that of man. They must lack a large field of stereoscopic vision because of the position of the eyes towards the sides of the head. Evidence such as we have points to the conclusion that ungulates do not have colour vision, and I can give no observations on the behaviour of deer which would lead me to refute it. They recognize perfectly well gradations between light and dark. I mentioned in the last chapter that the rutting stag wallows and darkens himself, and that deer are more afraid of human beings in dark suits than in light ones. It would be interesting to know why dark things frighten animals. My own experience is that white or very light-coloured collie dogs do not have the capacity for moving sheep which black or dark-coloured ones have. Their working movements, may be just as good but the sheep will not move with the same alacrity. This is well known among shepherds, and light-coloured pups are rarely kept.

The sight of deer is exceedingly acute to movement up to long distances, and in this way they are superior to most men. The observer may lie up and wait for deer approaching. This works well up to a point if his shape does not contrast with or alter the pattern of the landscape very markedly. But if the deer come within 60 or 70 yards he cannot keep still enough, for his eyes move. The eye and its movement are quite enough to shift deer. They do not like being looked at even if they are tame and will eat corn within a few yards of a human being. It is very striking to notice how a bunch of tame corn-fed deer tend to get behind you, and if you turn your head to them it is enough to move them. Similarly, tame deer do not like being photographed. The lens appears to have a similar quality to the eye in their vision.

Let us note the reactions of deer at a distance of 250 yards watching an observer walking along a footpath. At the first glimpse their heads are raised and the two forefeet come together (see the facing page). These two movements are synchronous. They stand motionless in whatever position they are and watch the person walking. As long as he keeps on walking all is well. They are not frightened and they do not move. But if the observer stops, probably they will go. If he makes any stealthy movements within their sight, such as crawling in the heather or getting behind a rock, they will certainly run away. Human beings in certain places, in certain postures, and gently moving on their way, are harmless creatures, it seems, but highly dangerous under other sets of conditions. It is quite easy, superficially, to explain this on the old association psychology, but the constant observer knows that there is something much deeper than this. Ninety per cent of deer never see a rifle or realize its significance, and of the other tenth few live to exercise their knowledge.

Deer do not like an observer to pass out of their sight if they are looking at him. If he should walk behind a knoll or through a dip it is most probable they will then run away. Another example

THE SENSES

a Hinds grazing   *b* They heard the click of the camera

Note the position of the fore-feet in a *and* b. *At gaze, the fore-feet are always brought together, giving the head all possible height.*

of the same principle is when an observer may be walking through the larger part of the edge of a circle with the deer as centre. The deer stand still and their heads turn as the observer moves, but there comes a moment when they can turn their heads no more, and unless they move their bodies they must turn their heads back again and look from the other side. That is the moment when the deer take fright because for that very short period of time they lose sight of the object of their attention. When a hind has spotted a stalker she can watch him for a long time. It is not at all unusual, if the situation is timed, to find that she will gaze steadily for half an hour, which is a long period for an animal to maintain concentrated attention. I have noticed, when close to groups of deer which have been suspicious of my presence, that there is interchange of 'sentry duty' but that young stags in a hind group do not often do this work. Here is an instance within one family of my feeding hinds, watched at the head of Glac Cheann in June 1935. I stalked to within 10 yards of them, which was too near, for they sensed me and moved to 50 yards *up wind* and started grazing again. The hind looked up and round occasionally for 20 minutes; then, with no sign apparent to me, it was obvious that the two-year-old stag was doing the watching. After a quarter of an hour the yearling hind (which was still sucking its mother) took the watch for a similar period. Then the hind herself began to take more interest. She watched me for five minutes, moved a few steps, and moved her neck and head from side to side. I knew I was identified. She barked 47 times at five- to 10-second intervals. Yet she did not go away, and although the youngsters took notice of me and the surroundings they did not seem startled. Eventually the hind walked away with that curious stilted walk, the reason for which I can never understand. She went for 100 yards and the young ones followed, gambolling with each other.

The action of a deer in moving a few steps or moving the head and neck to take another view is caused, presumably, by a desire to get a visual impression from another angle, owing to the fact that the area of stereoscopic vision in the visual field is small. By this means the valent object takes on depth and shape and the animal acts accordingly.

Deer do not tend in movement to pass below a human observer, and they will not allow him to get into a position above them. The hinds I fed in Glac Cheann did, towards the end of the time, allow me to stand above them in full view occasionally. The fact remains, however, that given the wind it is very much easier to approach deer from above. If we say they least expect molestation from that angle we may be reading too much into their minds, but they do not turn to look round above them in the same way that they scan the country below. The space above them seems largely outside their field of vision.

I have not seen deer look at dead members of their species with any sign of recognition of what they are or were, except once perhaps in July 1935, when I saw three hinds looking intently at something on the ground a few yards away. They pricked their ears, raised a forefoot, and walked round the thing in a circle. I picked up the object with my glass and found it to be a newly born calf. There were between 60 and 70 hinds in the group and within 10 minutes the attention of the whole herd was centred on this calf. They did not approach to within five yards of it. Then, minute by minute, hinds moved away, their interest lost. The calf had been born dead.

Touch

We can know little of this sense in ungulates, but it would seem that the muzzle is used as an organ of touch. The hinds are constantly touching their calves quite apart from licking them. Occasionally, I have seen a hind who has been suspicious of my presence lower her head to the ground. She has not done this to graze, and by watching her through the glass it has been possible to see that the hind has put her muzzle to the ground as if to *feel* any approach. I have watched this action very carefully; the jaws have been still and the muzzle has been pressed quite close to the grass. Some other stalkers have watched similar behaviour, and it would seem to be a normal method of sensing the approach of other heavy animals.

Taste

Deer show a catholicity in taste, but there are decided preferences. It should be recognized, however, that seasonal growth is also concerned and this will cause shift to a certain type of herbage irrespective of taste. The animals acquire appetites for some plants not normally found in the herbage of the hills. Turnips and potatoes are amongst these and tend to make deer the enemies of farmers. Taste reflects physiological condition; thus the need for minerals in the system, chiefly calcium and phosphorus, is felt as an appetite and is satisfied through the sense of taste. The deer of western forests are avid for calcium phosphate and they eat all cast antlers and bones. It is remarkable how soon a dead deer disappears from the ground. If it dies in April the greater black-back gulls and the hoodie crows are the first visitors to the carcass. They peck out the eyes and tongue and tear away the muzzle. A raven or an eagle may come next with stronger beaks and open the belly. If not, the hoodies have to wait awhile until putrefaction makes the skin more tender. Then they set to vigorously and clear the ribs and entrails. The sexton beetles are at work from below, removing an appreciable quantity of meat from the bones. Small muscid flies breed in the mass which filled the rumen, and soon the bones are bleaching in the sun, perfectly cleaned. The bluebottles are too late on the scene and get little chance of breeding on the dead deer of April. When the bones are white the deer begin to chew them; the ribs first, then the long bones and the fore part of the skull. Next spring only a bright green patch and perhaps the hooves remain to tell of the tragedy of last year.

This short account of the senses of red deer is quite inadequate and shows the limitations of observation. It is also some reflection of our total knowledge of the senses of animals. I have emphasized throughout the chapter the reactions of the deer through their sensory discrimination to the presence of the observer. This method provides us with concrete situations and vivid responses. Sometime in the future I hope to test the visual and aural discrimination of red deer if I can rear some tame hinds. These animals, by their quick responses, their capacity for protracted attention, and their expressive quality to those who know the species well, are excellent subjects for psychological experimentation.

CHAPTER ELEVEN

Conclusion

THIS BOOK IS not an attempt to propound a theory of animal behaviour, for as yet I do not have one. That is one of the reasons why I undertook the work described here. At the present stage of my growth I reject the behaviourist and mechanistic schools of thought as ultimately empty and illogical. Vitalism is equally illogical, involving as it does another dualism little different from that of the mechanists – 'materialism plus an entelechy' as Russell (1930) has said. The organismal point of view as set out by Ritter (1919) and later developed by Russell (1930, 1933, 1934, 1935) with particular reference to animal behaviour appeals to me strongly. The concept of organism recognizes the dimension of a living thing in time as well as space. The organism is a functional unity related both to its past and its future, and at any one moment presents but a phase of its life-history. Behaviour, as a function of organism, has a unifiedness and co-ordination bearing on the three main ends of growth, maintenance, and reproduction. As Russell (1924) says:

> The living thing is not a machine, for it shows persistent and prospective tendency or striving, and its responses are adjustable to a wide range of circumstances. It is not a mechanism accentuated by a psyche, for this formulation simply pushes the mystery further back, by ascribing to an immaterial agent or element of nature the faculties and powers of the organism as a whole. The living thing is not the clay moulded by the potter, nor the harp played upon by the musician. It is the clay modelling itself, or as Aristotle puts it, in a beautiful figure, being moulded directly by Nature herself, without the aid of tools 'but, as it were, with her own hands'.

The organismal view in biology has lighted the path along which

I have found myself groping. I see an animal and I want to know what it is doing, what it has been doing, and what it is likely to be doing soon. These are my first interests. As an individual I am much influenced by environment and I know the animal is affected much more. I wish to know more of this great bionomical complex concentric to the organism and at that point only do I become interested in morphology. Animal ecology and genetics, therefore, are the fields of science in which I have found myself working, but there is still some unifying principle missing. If I reach into comparative psychology I find myself in a bees' nest where methods and terminology are foreign to me. The organismal point of view provides a foundation upon which I can build, and I quote Russell (1930) once more:

> Organismal biology commences at the other end from biochemistry – with the study of the whole actions of organic unities, and it works down from the organism to its constituent parts, not upwards from the parts to the whole. It makes use of course of the analytical method, but bears always in mind the need to re-integrate the parts into the whole. It retains its independent status as the study of modes of action of organic unities considered in relation to their conditioning factors (action of lower unities, i.e., organic parts and physico-chemical processes) and to the main ends of the organism (development, maintenance, and reproduction). Biochemistry continues downward the study of the conditioning factors and the two methods dovetail into each other.
>
> On the other hand, at the opposite end of the scale, organismal biology merges easily into comparative psychology, as soon as the modes of action of the whole reach such a level as to give clear evidence of having reference to a perceived environment, when the organism must be regarded not merely as organism, but as percipient organism. Such is the case certainly with the higher insects and vertebrates and probably also much farther down the scale of beings. Organismal biology therefore appears to fit comfortably in between the psychological sciences on the one hand and the

physical on the other... We cannot claim for organismal biology anything like complete adequacy, or a close approach to full understanding of the living thing. The full secret of life will always elude a purely scientific treatment; it may be experienced, imagined, and felt, but never completely pinned down and explained. Something will always escape definition and measurement. Nevertheless we may rightly claim that the organismal method gives us a biology less remote from the truth than the abstract and schematic account to which the materialistic assumptions would limit us. It gives us a unitary biology, in which the abstractness and excessive analysis of the materialistic method are avoided; it allows us to look upon the living thing as a functional unity, disregarding the separation of matter and mind, and to realize how all its activities – activities of the whole, and the activities of the parts, right down to intra-cellular unities – subserve in cooperation with one another the primary ends of development, maintenance, and reproduction.

The great bulk of papers on animal behaviour lift the organism from its normal environment and place it in a set of artificial conditions, and this often results in findings which are not valid for interpretation of representative behaviour. [I would mention as notable exceptions the work of Howard (1907–15, 1920, 1929) on birds, and Huxley's paper (1914) on the great crested grebe.] The preliminary studies of animals in their natural surroundings, which appear to me as the initial steps to any experimental approach, are, unfortunately, of rare occurrence. It happens that such studies take much time and patience and frequently isolate the observer from intellectual contact with his fellows. They are not tasks for the laboratory and daily common-room discussion.

We may take it as axiomatic that an animal strives to keep itself within an ecological norm. The behaviour which results is characterized by the following qualities which have been set out and discussed fully by McDougall in his *Outline of Psychology*. The marks of behaviour are:

1 A certain spontaneity of movement.
2 Persistence of activity independently of the continuance of the impression which may have initiated it.
3 Variation of direction of persistent movement.
4 Termination of the animal's movements as soon as they have brought about a particular kind of change in its situation.
5 Preparation for the new situation toward the production of which the action contributes.
6 Some degree of improvement in the effectiveness of behaviour, when it is repeated by the animal under similar circumstances.
7 Purposive action is a total action of the organism (as opposed to reflex action, which is always partial).

Within the ecological norm, the environment exerts pressures of one kind and another and behaviour is influenced. The animal adapts itself to preserve the safety of the organism and, as Lundholm (1934) has pointed out, there are two general kinds of adaptive processes – adaptation by deference to the environment, and by defiance or control of it. Deer, in their behaviour, usually defer to the changing environment by movement, but the lemming of the tundra, in its small way, defies it by creating its own world of a higher temperature under the snow.

But environment is a very wide term, embracing the animal's fellows and the social system in which it lives. Our study becomes socio-ecological and we find a social system which, in some measure, is a control of the environment. There is this large and interesting field between zoology and psychology which few workers seem willing to explore. If we are to watch one of the higher animals and measure, as accurately as we know how, the environmental influences on behaviour, the subject for study preferably should exhibit marked reactions; and if, as they should, our observations are to extend over a long period, it is better for us that the animal should live above ground. When the list of British mammals is considered, very few species meet this latter desirable requirement of our studies. I have remarked that animals which live above ground adapt themselves to the environment by movement; they do not evade it by subterranean habits which we cannot observe adequately. Where a species is of social habit, I would emphasize the necessity of taking sociality fully into account in observing and interpreting behaviour. The life-history of the red deer would be an empty and meaningless thing divorced from the sociality which is the very foundation of their existence.

This book tries to give the plain tale of an animal's life, of the things it does and is trying to do, of its relations with its fellows and with men, and of the things to which, as long observation leads me to believe, it responds. The study can be considered in no way complete, for now, after two years, I am left with more problems to solve than I knew of at the outset.

My estimate of animal mind from long contact with it is high, and these two years of intensive observation in critical vein have not lowered it. Lloyd Morgan's maxim (quoted, page 86) in studying animal behaviour is a good one, but there is no need to set up artificial standards of simplicity or for one school of thought to impose its own criteria as being the ineluctable measure of simplicity. I am not entirely content to accept the evidence of anatomical science as being final when it is called in to show that the animal's brain, where present, and its sense organs are of such a nature to preclude certain kinds of experience. Quality of work is not to be inferred from the up-to-dateness of the workmen's tools. May I take an analogy from the genetics of *Drosophila*? One race of mutants, 'white eye', is indistinguishable anatomically from the normal wild type. They can be discovered only by breeding tests. Although they are homozygous, or pure-breeding, for the 'white eye' factor, the organism has in course of time become outwardly adjusted to the deficiency and the eye appears normal. Something has been achieved for which the materials might seem not to be present. The organism in the dimension of time has remarkable elasticity. And so with animal structure and behaviour. The behaviour of one species can show surprising latitude under stress of circumstance, and amongst the higher animals we find response to sets of conditions and a spontaneity in action which we, as so-called rational beings, could not better. In some instances I feel that the most simple explanation of an act of behaviour is to follow the bare outline of our own mental processes in such a situation. I believe the teleological approach to animal behaviour to be dangerous, but the current objection to anthropomorphism can be overdone. Who are the people with whom the higher animals are most serene, and who achieve most success in their management and training? Not those who look upon them as automata, but those who treat them as likeable children of our own kind.

Glossary of Gaelic Place-names

A' MHAIGHDEAN	The maiden.
ACHACHIE	The field of little fields.
ALLT AIRDEASAIDH	The water of the point of the waterfalls.
ALLT CREAG ODHAR	The water of the dun-coloured rock.
AN POLAIN	The hollow of soft ground.
AN TEALLACH	The hearth, *or* the forge.
BEINN A' CHAISGEIN BEAG	The small hill of cheese.
BEINN A' CHAISGEIN MOR	The big hill of cheese.
BEINN A' CHLAIDHEIMH	The sword hill.
BEINN DEARG BHEAG	The small red hill.
BEINN DEARG MHOR	The big red hill.
BIDEIN A' GHLAS THUILL	The sharp-pointed peak of the grey hole.
CAISEAMHEALL	The rounded hill of cheese.
CARN AIRIDH AN EASAIN	The cairn of the sheiling by the waterfalls.
CARN NA BEISTE	The cairn of the beast (probably bear).
CARN NAM BUAILTEAN	The cairn of the purple stork's-bill flower.
CARN NA CARNACH	The cairn of many cairns.
CARN NA-H-AIRE	The cairn of the sheiling.
COIR' A' GHAMHNA	The corrie of the stirks.
COIR' A' GHIUBHSACHAIN	The corrie of the pine trees.
COIR' A' MHUILLIN	The mill corrie.
COIRE MOR AN TEALLAICH	The great corrie of the Teallach.
CREAG-MHEALL BEAG	The small, rocky, rounded hill.
CREAG-MHEALL MEADHONACH	The middle, rocky, rounded hill.
CREAG-MHEALL MOR	The big, rocky, rounded hill.
EAS BAN	The white fall.
FIONN LOCH	The white loch.
GARBH ALLT	The rough water.
GLAC CHEANN	The hollow of the heads.

GLAS MHEALL BEAG	The small, grey, rounded hill.
GLAS MHEALL MOR	The big, grey, rounded hill.
GLAS THUILL	The grey hole (corrie).
GLEANN AN NID	The glen of the nest.
GLEANN CHAORACHAIN	The glen of the sheep.
LARACHANTIVORE (LARACH AN TIGH MHOR)	The foundations of the big house.
LOCHAN A' BEARTA	The little loch of the loom.
LOCH AN EICH DHUIBH	The loch of the black horse.
LOCHAN GAINEAMHAICH	The little sandy loch.
LOCH A' MHADAIDH MOR	The loch of the big wolf.
LOCH GHIUBHSACHAIN	The loch of the pine trees.
LOCH MOR BAD AN DUCHARAICH	The big loch of the clump of willows.
LOCH TOLL AN LOCHAIN	The lochan of the hole.
MAC' US MATHAIR	Son and mother.
MEALL BHUIDHE	The yellow rounded hill.
RUIGH MHEALLAN	The rounded hills with grassy places.
SAIL BHEAG	The little heel.
SAIL LIATH	The grey heel.
SAIL MHOR	The great heel.
SGURR FIONA (FHEOIN)	The peaked, rocky hill of wine.
SGURR NAN EICH	The peaked, rocky hill of the horse.
SGURR RUADH	The red peaked, rocky hill.
SHENAVALL (SEANA BHAILE)	The old town.
STRATH NA SHEALLAG (SEALGAIR)	The wide glen of the hunter.
UISGE TOLL A' MHADAIDH	The water of the wolf's hole.

Bibliography

Allee, W. C. 1932. *Animal Life and Social Growth*. Baltimore.

Alverdes, F. R. 1927. *Social Life in the Animal World*. London.

> The author has collected data of sociality in many species of animals from a wide variety of sources.

Baker, J. R., and Ranson, R. M. 1932. 'Factors affecting the Breeding of the Field Mouse (*Misrotus agrestis*). I. Light.' *Proc. Roy. Soc. B*, 110, 313–22.

> The authors found that the shortening of the daily period of exposure to light from 15 hours to nine hours almost prevents reproduction in the field vole. The female is chiefly affected.

– 1933. 'Factors affecting the Breeding of the Field Mouse (*Microtus agrestis*). III. Locality.' *Proc. Roy. Soc. B*, 113, 486–95.

> There exists a general correlation between hours of sunshine per month and breeding condition of the voles.

Bentham, G., and Hooker, J. D. 1887. *Handbook of the British Flora*, 5th edition. London.

Bierens De Haan, J. A. 1929. *Animal Psychology for Biologists*. London.

> Lectures given in the University of London. The author stresses the response of animals to complexes rather than to single features in the perceptual field.

Bissonnette, T. H. 1930. 'Studies on the Sexual Cycle in Birds. I. Sexual Maturity, its modification and possible control in the European starling (*Sturnus vulgaris*).' *Amer. J. Anat.* 45, 289–302.

– 1931. 'Studies on the Sexual Cycle in Birds. V. Effects of white light of different intensities upon the testis activities of the European starling (*Sturnus vulgaris*).' *Physiol. Zool.* 4, 542–74.

- 1933. 'Light and Sexual Cycles in Starlings and Ferrets.' *Quart. Rev. Biol.* 8, 201–8.
- 1935. 'Relations of Hair Cycles in Ferrets to Changes in the Anterior Hypophysis and to Light Cycles.' *Anat. Rec.* 63, 159–68.

    There is a positive correlation between hair growth and hypophyseal activity.

- 1935. 'Modification of Mammalian Sexual Cycles. III. Reversal of the cycle in male ferrets (*Putorius vulgaris*) by increasing periods of exposure to light between 2 October 2 and 30 March.' *J. Exp. Zool.* 71, 341–73.

Cameron, A. E. 1932 (a). 'The Nasal Bot Fly, *Cephenomyia auribarbis* Meigen (Diptera, Tachinidae) of the Red Deer, *Cervus elaphus* L.' *Parasitology*, 24, 185–95.

- 1932 (b). 'Arthropod Parasites of the Red Deer (*Cervus elaphus* L.) in Scotland.' *Proc. Roy. Phys. Soc.* 22, 81–9.
- 1934. 'The Life History and Structure of *Haematopota pluvialis*, Linné (Tabanidae).' *Trans. Ray. Soc. Edin.* 58, 211–50.

Cameron, A. G. 1923. *The Wild Red Deer of Scotland*. Edinburgh.

    Gives an account of Evans's work during 25 years' occupancy of the island forest of Jura. A most interesting book. There are chapters on antler form.

Cameron, T. W. M. 1932. 'Some Notes on the Parasitic Worms of the Scottish Red Deer.' *Proc. Roy. Phys. Soc.* 22, 91–7.

Cameron, T. W. M., and Parnell, I. W. 1933. 'The Internal Parasites of Land Mammals in Scotland.' *Proc. Roy. Phys. Soc.* 22, 133–54.

Crew, F. A. E. 1927. 'The Scrotum: A Temperature-regulating Mechanism.' *Verhandl. internat. Kong. f. Sexualforschung. Experimentalforschung u. Biologie*, 72–85. Berlin.

Davies, Glyn. 1931. 'An Unusual Sex-Ratio in Red Deer.' *Nature*, 127, 94.

Elton, C. 1927. *Animal Ecology*. London.

    A foundation work.

– 1932. 'Territory among Wood Ants (*Formica rufa* L.) at Picket Hill.' *J. Anim. Ecol.* 1, 69–76.

Fitch, W. H. and Smith, W. G. 1887. *Illustrations of the British Flora*, 2nd edition. London.

Geddes, P., and Thomson, J. A. 1901. *The Evolution of Sex*, revised edition. Contemporary Science Series.

Grant, R. 1934. 'Studies on the Physiology of Reproduction in the Ewe. I. Symptoms, Periodicity and Duration of Oestrus.' *Trans. Ray. Soc. Edin.* 58, 1–15.

Grimshaw, P. H. 1894. 'On the Occurrence in Ross-shire of *Cephenomyia (rufibarbis) auribarbis*, a new British Bot Fly parasitic on the Red Deer.' *Ann. Scot. Nat. Hist.* 4, 155–8.

Groos, K. 1898. *Play of Animals*.

Hadwen, S., and Palmer, L. J. 1922. *Reindeer in Alaska*. US Dept. Agric. Bull. No. 1089.

Heape, W. 1913. *Sex Antagonism*. London.

– 1931. *Emigration, Migration and Nomadism*. Cambridge.

The author has collected a wealth of data, and his thesis is, that mass emigrations in certain species such as lemmings and springbok are referable to periodical surpluses of 'generative ferment' which we now identify with anterior pituitary hormone.

Hingston, R. W. G. 1933. 'Psychological Weapons in Animal Fight.' *Character and Personality*, 2, 3–21.

Hobhouse, L. T. 1913. *Development and Purpose: an Essay towards a Philosophy of Evolution*. London.

– 1915. *Mind in Evolution*, 2nd edition. London.

The author ably criticizes Thorndike's mechanistic theories of animal behaviour, and recounts experiments of his own on 'insight' learning.

Howard, H. Eliot. 1907–14. *The British Warblers: A History with Problems of their Lives*. London.

An account of the behaviour of these birds, especially at breeding time and in relation to territories.

– 1920. *Territory in Bird Life.* London.
– 1929. *An Introduction to the Study of Bird Behaviour.* Cambridge.

> An extension and development of views in the previous work and observations on the differential time threshold in mating behaviour in several species.

Huxley, J. S. 1914. 'The Courtship-habits of the Great Crested Grebe (*Podiceps cristatus*): with an Addition to the Theory of Sexual Selection.' *Proc. Zool. Soc. Lond.* 491–562.

– 1931. 'The Relative Size of Antlers of Deer.' *Proc. Zool. Soc. Lond.* 819–64.

Keith, A. 1916. 'On Certain Factors concerned in the Evolution of the Human Race.' Presidential Address to the Royal Anthropological Institute. *J. Roy. Anthropol. Inst.* 66.

Klemola, V. 1929. 'On Breeding and Distribution of Reindeer in Eurasia' [translated title]. *Terra*, 11, 137–63. (Finnish with German summary.)

Köhler, W. 1925. *The Mentality of Apes.* London.

> A classic of animal behaviour. The author is a prominent 'Gestalt' psychologist and in this work gives many examples of 'insight' learning.

Kropotkin, P. 1904. *Mutual Aid: a Factor of Evolution*, revised edition. London.

> The first part of this work gives many instances of mutual aid within and between certain species of animals. The author was not very critical in collecting his data and his outlook was teleological in a marked degree.

Leopold, A. 1933. *Game Management.* New York.

> Applied ecology. A very valuable work.

Lillie, R. S. 1932. *Sex and Internal Secretions.* Chicago.

Lundholm, H. 1934. *Conation and Our Conscious Life.* Duke Univ. Psychol. Monographs.

McDougall, W. 1931. *Social Psychology*, 22nd edition. London.

McDougall, K. D., and McDougall, W. 1931 (b). 'Insight and Foresight in Various Animals – Monkey, Racoon, Rat and Wasp.' *J. Comp. Psychology*, 11, 237–73.

McDougall, W. 1933. *An Outline of Psychology*, 6th edition. London.

McKenzie, F. F., and Phillips, R. W. 1933. 'The Effect of Temperature and Diet on the Onset of the Breeding Season (Estrus) in Sheep.' *Ann. Rpt. Missouri Agric. Exp. Sta. Bull.* No. 328, 13–14.

Millais, J. G. 1904–6. *The Mammals of Great Britain and Ireland*. London.

> The author gives those morphological details of red deer which I have purposely omitted as being irrelevant.

Miller, W. C. 1932. 'A Preliminary Note upon the Sex Ratio of Scottish Red Deer.' *Proc. Roy. Phys. Soc.* 22, 99–101.

Morgan, C. Lloyd. 1894. *Introduction to Comparative Psychology*. London.

Nichols, J. E. 1927. 'Meteorological Factors affecting Fertility in the Sheep.' *Zeit. indukt. Abst. u. Vererb.* 43, 313–29.

Palmer, L. J. 1926. *Progress of Reindeer Grazing Investigations in Alaska*. US Dept. Agric. Bull. No. 1423.

Portland, His Grace The Duke Of. 1933. *Fifty Years and More of Sport in Scotland*. London.

Rasmussen, A. T. 1917. 'Seasonal Changes in the Interstitial Cells of the Woodchuck (*Marmota monax*).' *Am. J. Anat.* 22, 475.

Ritter, W. E. 1919. *The Unity of the Organism, or the Organismal Conception of Life*. Boston.

Rowan, W. 1931. *The Riddle of Migration*. Baltimore.

> A popular account of the author's work on the effect of light on the gonads and consequently on migration. Bissonnette (q.v.) has extended and developed Rowan's views.

Russell, E. S. 1924. *Study of Living Things*. London.

– 1930. *The Interpretation of Development and Heredity. A Study in Biological Method*. Oxford.

- 1933. 'Is Comparative Psychology an "Objective" Science?' *Scientia*, 181–90.
- 1934. *The Behaviour of Animals.* London.

  Lectures given in the University of London. The author emphasizes the organismal point of view and the value of natural history studies of animals. A work of fundamental importance.
- 1934. 'The Study of Behaviour.' Presidential Address, Section D, Brit. Assoc., Aberdeen.
- 1935. 'Valence and Attention in Animal Behaviour.' *Acta Biotheoretica*, A. 1, 91–9.

Scrope, W. 1894. *Days of Deer Stalking.* London.

A classic of the sport, and full of anecdotes of the habits of deer.

Sdobnikov, V. M. 1935. 'Relations between the Reindeer (*Rangifer tarandus*) and the Animal Life of Tundra and Forest' [translated title]. *Trans. Arctic Inst., Leningrad*, 24, 5–60. English summary, 61–6.

Slonaker, J. R. 1924. 'The Effect of Pubescence, Oestruation, and Menopause on Voluntary Activity in the Albino Rat.' *Amer. J. Physiol.* 81, 325.

Spiker, C. J. 1933. 'Some Late Winter and Spring Observations on the White-Tailed Deer of the Adirondacks.' *Roosevelt Wild Life Bull.* 6, 327–85.

Stuart, J. S., and Stuart, C. E. 1848. *Lays of the Deer Forest.* Edinburgh.

The second volume of this work contains many personal observations of a detailed nature on the habits of deer.

Tansley, A. G. 1911 (Editor). *Types of British Vegetation.* Cambridge.

Thomson, J. A., and Geddes, P. 1931. Life: *Outlines of General Biology.* London.

Townsend, M. T., and Smith, M. W. 1933. 'White-Tailed Deer of the Adirondacks.' *Roosevelt Wild Life Bull.* 6, 153–325.

Turner, C. W. 1934. *The Causes of the Growth and Function of the Udder of Cattle.* Missouri Agric. Exp. Sta. Bull. No. 339.

Washburn, M. F. 1930. *The Animal Mind*, 3rd edition. New York.

    An important book gathering together and discussing the results of much of the work done in the last 30 years. There is a full bibliography.

Wheeler, W. M. 1928. T*he Social Insects: Their Origin and Evolution*. London.

Wiesner, B. P., and Sheard, N. M. 1933. *Maternal Behaviour in the Rat*. Edinburgh.

    A striking example of the application of endocrine physiology to the study of one aspect of animal behaviour.

Index

A
Acreage: and density, of ground worked, 36
Age classes, 41
  mortality of, 42, 43, 44, 75,
Age of deer, 40
Allee, W. C., 64, 199
Altitude: and movement, 101
  and snow, 119, 121
  and temperature, 97
Anterior pituitary, 66
  and antlers, 160
  stimulation of, by light, 151
Antlers: and anterior pituitary, 160
  cast earlier on gneiss, 12
  and fighting, 154–160, 168–169
  and reproduction, 153–157
  and sex ratio, 37
  shedding dates, 161
  types of, 93
  'velvet', 62, 155

B
Baker, J. R., 151, 199
Barometric pressure, vi, 108
Behaviour, McDougall's marks of, 193
Bentham, G., 148, 199
Birds: blackbird, 5
  black-throated diver, 11
  buzzard, 5
  Canadian crow, 151
  crested tit, 6
  cuckoo, 5
  dipper, 2, 8
  fieldfare, 2, 5
  goldcrest, 6
  golden eagle, 1, 42
  goosander, 11
  great black-backed gull, 189
  greenshank, 11
  grey wagtail, 6
  grouse, 42, 185
  hooded crow, 136, 189
  junco finch, 151
  kestrel, 5
  meadow pipit, 5
  ptarmigan, 1
  raven, 189
  redstart, 5
  redwing, 5
  ring ouzel, 5, 11
  robin, 5
  rock pipit, 5
  snipe, 9
  snow bunting, 1
  starling, 151–152, 199–200
  tits, 5–6
  wheatear, 5
  whooper swan, 11
  willow warbler, 5
  wren, 5–6
Bissonnette, T. H., 151, 199, 203
Bogwood, 13, 60
Bot fly, 136
Boundaries, 29–30, 35, 49, 52, 119, 167
  discreteness of, as ascertained by feeding experiment, 57
  see Territory.

C
Calving time, 71
  and game policy, 38
  the hind at, 176

Cameron, A. E., 126, 134, 136–137, 141, 200
Cameron, A. G., 37, 40–41, 74–75, 200
Cameron, J. 42, 53, 109
Cameron, T. W. M., 141, 200
Cephenomyla auribarbis, 128, 134, 200
Chrysops relicta, 128, 130
Clegs, xxxiii, 126–132
Colonization, 49, 57, 63, 95
Colour, 21, 23, 159, 185
Corvus brachyrhynchos, 151
Courtship behaviour, 173–176
Cow, voice of, 80
Crew, F. A. E., 162, 200
Cripples: behaviour of, 77
  and game policy, 38–40
  in relation to rutting behaviour and antler growth, 160
Culicoides pulicaris, 128, 140
Curiosity, primordial, 172, 180

D
Davies, G., 44, 200
  Deer, see each species.
  Density of population, 30–32
  psychological factor, 34
Dog, 32, 68, 80, 85, 91, 149–150, 170, 184–185
Drosophila, 65, 195
Durkheim, E., 9

E
Eastern schist, undifferentiated, 4
Elton, C. S., 28, 200
Equipment: binoculars 19, 75, 135
  camera and lenses, xxxii, 19
  Grenfell cloth, 18, 24
  Harris tweed, 18
  telescope, 15, 19, 53, 133
  tent, 18
  Tricouni nails, 18
  waterproofs, avoidance of, 18

F
Fallow deer, 120
Fertility ratio, 40, 43–44
Fighting, 28, 37, 153, 157, 159, 163, 168, 171
Fitch, W. H., 148, 201
Foresight, 85–87, 95
Frogs, early breeding, frost killing, and deer eating, 6; tadpoles, 64
Frost: and humidity, 116
  and movement, 103
  and sense of smell, 181–182
  travel in, 15
  and vegetation, 142

G
Gait, 51, 78, 158
Game policy and percentage of cripples, 38–40
Geddes, Sir P., 64, 73, 201
Geographical distribution, 27
Gestalt School, 85
Gonads and antlers, 151–157
  dual function of, 151
Gregariousness: comparison of hunters and grazers, 27
  and sexual jealousy, 88
Grimshaw, P. H., 134, 201
Groos, K., 82–83, 201

H
Habit, conservatism of, 27
Hadwen, S., 34, 43, 161, 201
Haematopota crassicornis, 126, 128
Haematopota pluvialis, 126, 128
Hail, 13, 14, 103, 105, 119, 175
Harem, 154, 157, 163, 166–167, 169, 170, 172–174
Heape, W., 73, 163, 201

INDEX

Hearing, 184–185
Heather burning, 54
Herbage: alpine compared with that on peat, 2
  associations, 143–148
  of Lewisian gneiss, 10–12
  and sense of taste, 189
  see Vegetation.
Hibernation, 152
Hinds: at calving time, 71, 176
  identification of, 22
  inquisitiveness of, 67, 184
  lactation, 23, 66, 177–178
  movement of, 67, 71, 132–133
  play of, 86 et seq. 82–87
  territories, 29–30, 47
  voice of, 80
  wanderings of,
Hingston, R. W. G., 158–159, 165, 201
Hippoboscidae, 128, 137
Hobhouse, L. T., 88, 201
Hooker, Sir J. D., 148, 199
Howard, E., 27, 193, 201
Humidity, 13, 15
  in frost and thaw, 116
  and movement, 95–96, 100, 103–105, 108–117
  and scent, 112–113, 181–182
  and snow, 121
Hummel stags, 153–154
Huxley, J. S., 154, 193, 202
Hypoderma bovis, 136
Hypoderma diana, 128
Hypophysis, see Anterior pituitary.

I

Imagination and natural scenery, 16
Insects, effect on movement, 126–149
Insight, 85
Ixodes ricinus, 128, 137

J

Jennings, H. S., 24
Junco hyemalis connectens, 151
Jura, 37, 41

K

Keds, 43, 136, 137, 139, 159
Keith, Sir A., 93–94, 202
Klemola, V., 43, 116, 202
Köhler, W., 73, 85, 202
Kropotkin, Prince, 120, 202

L

Lactation, 23, 66, 177–178
Larynx, 159
Leadership, 68 et seq. 65–72, 89, 91
Leopold, A., 34, 36–38, 100, 202
Lice, 137
Lillie, R. 5., 151, 202
Lipoptena cervi, 128, 136
Lizard, common, 5
Lundholm, H., 194, 202

M

MacDonald, D., 14, 75, 109
MacDonald, R., 177
McDougall, W., 83, 85–86, 193, 202–203
Mammals: antelope, 101
  apes, 73, 85
  bison, 91, (as buffalo, 101)
  caribou, 91, see Reindeer.
  cattle, 9, 33, 76, 80, 83, 136, 150
  dog, 32, 68, 80, 85, 91, 149, 150, 170, 184–185
  elephant, 76, 91
  fallow deer, 120
  ferret, 151
  fox, 8, 42, 88
  goat, 2, 89–90, 185
  hare, 184
  hedgehog, 5

house mouse, 150
lemming, 194, 201
long-tailed field mouse, 5, 199
mole, 5
mountain hare, 7
mountain sheep, 101
mule deer, 37, 101
otter, 5
ponies, 9, 62, 127, 129
rabbit, 4–5, 42, 58, 141–142
reindeer, see sep. entry.
roe deer, see sep. entry.
sheep, see sep. entry.
squirrel, 5, 110, 150
voles, 5, 150–151, 199
weasel, 184
wild cat, 5, 8, 42, 88
woodchuck, 152, 203
Map-distances, 35
Marmota monax, 152, 203
Masturbation, 157, 166
Maternal behaviour, 66, 71, 178
 in stampede, 132–133
Matriarchy, 29, 78, 88–89
Melophagus minus, 136
Microtus agrestis, 151, 199
Midges, 62, 99, 132, 140
Millais, J. G., 75, 93, 203
Miller, W., 44, 203
Mineral food: eating antlers, 155
 eating bones, 189–191
 eating dead frogs, 6
 newly burnt ground, 55
 peaty and peat-free ground, 2
Mist, 2, 14–16, 99, 114–115, 120, 163, 175
Moore, C., 162
Morgan, C. Lloyd, 86, 195, 203
Mortality, winter-, 39
 calf-, 86, 195, 203
Movement of deer, 16
 and altitude, 101

comparisons, 91
daily, 56, 75, 100–101, 106, 117–118, 123
effect of insects on, 126–149
and frost, 103–108, 181–182
and hail, 119
of hind groups, 67, 71, 132–133
and humidity, 95–96, 100, 103–105, 108–117
and light, 123
of milk hinds, 70
and rainfall, 118
seasonal, 31, 56–57, 60
and snow, 119–123
of stags in single file, 73
stampede, 132–133
trekking distances, 52, 91, 164–165
vegetation, 141–149
and weather, 95–125
and wind, 117–118
of yeld hinds, 70
Mule deer: daily range, 101
 sex ratio, 37

N
Newt, palmated, 3, 13
Nostril fly, 128, 134–135, 139

O
Oedemagena, 136
Organismal biology, 192–196

P
Palmer, L. J., 34, 43, 45, 161, 201, 203
Parasites, 43, 128, 134, 137–141, 200
Parnell, I. W., 141, 200
Paths made by deer, 58–60
Photography, 19, 186
Pituitary, anterior lobe of, see Anterior pituitary.

Play, 82–87
Population: distribution and density of, 21, 27, 31–45
  fertility ratio, 40
Portland, Duke of, 158, 203
Precipitation, see rainfall, snow, hail.

Q
Quartzite slabs, 3

R
Rainfall, 9, 14–15, 118
Range, and territory, 27
Red deer: age of, 40
  colour of, 21, 23, 159, 185
  copulation of, 172–173
  daily range, 100
  distinguishing classes of, 21–22
  distribution in Scotland, 26
  gait, 51, 78, 158
  movement, see sep. entry.
  paddocked, 33
  paths, 58–60
  play, 82–87
  rubbing against trees, 61–63
  senses, 179–190
  sex ratio, 37, 39, 43–44
  sociality, see sep. entry.
  voice, 80–82
  wallowing, 58–63, 103, 135–140, 158–159, 174
  as woodland animal, 50
  wounded, 38, 77, 161
Reindeer: antlers of, 140, 161
  fertility ratio, 43
  grazing density, 34
  and insects, 129, 136, 140
  social system, 92
  and thaw, 116
  trekking of caribou, 91
Reproduction, 150–178 see Anterior pituitary, Antlers, Gonads, Movement, Territory, &c.
Ritter, W. E., 191, 203
Roe deer: 51
  in Dundonnell, 5
  gait, 51
  grazing density, 34
  and insects, 140
  social system, 88
  twinning and calf mortality, 43
Rowan, W., 151, 203
Russell, E. S., 191–192, 203
Rutting behaviour in relation to territories, 29–31, 166–176

S
Scent: and territory, 28
  and humidity, 112–115
Schweitzer, A., 64
Sdobnikov, V. M., 92, 129, 136, 140, 204
Senses, 179–190, see sep. entries.
Senses of the observer, 21–24
Sentry duty, 187
Sex ratio: differential, 37
  mule deer, 37
  primary, 43
  secondary, 44
Sexton beetles, 190
Sheard, N., 66, 178, 205
Sheep: Blackface, xvii, xxvi, 90
  on deer ground, 32, 35, 51–52
  flocking instinct, 90–91
  Merino, 90, 92
  Shetland, 91
  Soay, 91–92
  social system, 90–91
  sound of when alarmed, 185
  temperature and oestrus, 163, 108
  and ticks, 139
Sight, 185–189
  and reading, 21

Slonaker, J. R., 204
Smell: in relation to humidity, 112–113, 181–182
  sense of, 171, 180–184
Smith, M. W., 34, 38, 43, 204
Smith, W. G., 148, 201
Snow: demonstrating air movements, 2
  and movement of deer, 119–123
  and senses, 182, 184
  sound and east wind, 10
  travelling on, 15
  and vegetation, 142
Sociality: advantages of, 64
  lactation and, 66
  rearrangement of groups, 70–72
  reproductive processes and, 66, 150–178
  and the rut, 166–176
  sexual differences in, 67
  stag companies, 72–78
Social system, 29, 64–79
  goat, 89
  hind group, 65–72, 92
  reindeer, 92
  roe deer, 87–88
  sheep, 90
Socio-ecology, 28
Spencer, H., 82
Stags: antlers, see sep. entry.
  breeding season, 30, 152–153
  on burnt ground, 55
  colour at rut, 159, 185
  companies, 72–78
  fighting, 28, 37, 153, 157, 159, 163, 168, 171
  gait, 158
  hummels, 153–154
  larynx, 159
  in old age, 76
  social behaviour of young, 67, 176

  travelling, 158, 164
  voice, 81, 159, 164
Stalking, technique, 20–25
Stampede, 68, 115–116, 127, 130, 132–133
Stereoscopic vision, 185, 188
Sunshine: annual curve, 13–14
  and behaviour of deer, 123
Sweat glands, 162
Switch horns, 154

T
Tabanid flies, 53, 126, 129, 132, 134
  Tabanus distinguendus, 128
  Tabanus montanus, 128
  Tabanus suedeticus, 128
Tansley, A. G., 148, 204
Taste, 189–190
Temperature: annual curve, 13
  of hibernating animals, 152
  and morning mist, 16
  and movement, 97–108
  recuperative effect of cold, 164
  and rut, 162–164
Territory: boundaries, 47–58
  breeding or rutting, 30, 167–178
  hinds', 29, 30, 47
  neutral ground at rutting season, 164
  range of, 26–46
  sociality and, 47–64
  stags', 47–58
  summer, 29–30
  trekking and, 91
  winter, 29
Testes: and antlers, 160
  and hibernation, 152
Testis hormone, 159–160
Thaw: effect of, 15
  and herd movements, 121–123
  and humidity, 116

and vegetation, 142
Thomson, Sir J. A., 201, 204
Ticks, 43, 124, 139, 177
Touch, 189
Townsend, M. T., 34, 38, 43, 204
Trees, rubbing, v, 58, 61–63
Tundra, 92, 129, 136, 140, 194, 204
Turner, C. W., 66, 178, 204
Twins, 43

V

Vegetation and movement, 141–149
  see Herbage.
Virginia deer, 34
Vision, see Sight.
Voice, 80–81, 151, 159

W

Wallows, 58, 60–61, 63, 136–137, 158, 185
Weather: general description, 13
  effect of on behaviour, see Rainfall, Snow, Wind, Temperature, Humidity, &c.
Wheeler, W. M., 28, 205
White-tail deer: fertility ratio, 43
  grazing density, 36
  sex ratio, 38
Wiesner, B. P., 66, 178, 205
Wind: in corries, 2
  and movement of deer, 117–118

Luath Press Limited

committed to publishing well written books worth reading

LUATH PRESS takes its name from Robert Burns, whose little collie Luath (*Gael.*, swift or nimble) tripped up Jean Armour at a wedding and gave him the chance to speak to the woman who was to be his wife and the abiding love of his life. Burns called one of 'The Twa Dogs' Luath after Cuchullin's hunting dog in Ossian's *Fingal*. Luath Press was established in 1981 in the heart of Burns country, and now resides a few steps up the road from Burns' first lodgings on Edinburgh's Royal Mile.

Luath offers you distinctive writing with a hint of unexpected pleasures.

Most bookshops in the UK, the US, Canada, Australia, New Zealand and parts of Europe either carry our books in stock or can order them for you. To order direct from us, please send a £sterling cheque, postal order, international money order or your credit card details (number, address of cardholder and expiry date) to us at the address below. Please add post and packing as follows: UK – £1.00 per delivery address; overseas surface mail – £2.50 per delivery address; overseas airmail – £3.50 for the first book to each delivery address, plus £1.00 for each additional book by airmail to the same address. If your order is a gift, we will happily enclose your card or message at no extra charge.

Luath Press Limited
543/2 Castlehill
The Royal Mile
Edinburgh EH1 2ND
Scotland
Telephone: 0131 225 4326 (24 hours)
Fax: 0131 225 4324
email: sales@luath.co.uk
Website: www.luath.co.uk